3D PRINTING

3D PRINTING

A REVOLUTIONARY PROCESS FOR INDUSTRY APPLICATIONS

RICHARD SHENG
Jiaotong University, Shanghai, China
Webster University, Shanghai, China
SAE International, Shanghai, China

ELSEVIER
ACADEMIC PRESS
An imprint of Elsevier

Academic Press is an imprint of Elsevier
125 London Wall, London EC2Y 5AS, United Kingdom
525 B Street, Suite 1650, San Diego, CA 92101, United States
50 Hampshire Street, 5th Floor, Cambridge, MA 02139, United States
The Boulevard, Langford Lane, Kidlington, Oxford OX5 1GB, United Kingdom

ISBN: 978-0-323-99463-7

For information on all Academic Press publications visit our website at https://www.elsevier.com/books-and-journals

Publisher: Matthew Deans
Acquisitions Editor: Glyn Jones
Editorial Project Manager: Naomi Robertson
Production Project Manager: Sojan P. Pazhayattil
Cover Designer: Greg Harris

Typeset by TNQ Technologies

Contents

Preface

The main purpose of this book is to introduce the importance of 3-D printing/additive manufacturing and how this technology is impacting our daily lives. 3-D printing/additive manufacturing is one of the most revolutionary inventions of the 21st century. The 3-D printer is a fascinating piece of equipment as it prints real-life three-dimensional objects instead of just printing some words or images on a piece of paper. 3-D printing/additive manufacturing makes solid objects from a digital file. An object is printed using an additive process, which creates an object by laying down layers of materials until the object is finally created. A 3-D printer produces complex shapes while using less material as compared to traditional manufacturing processes. 3-D printers can be used to print any object, irrespective of their material. Nowadays, through technical advancements, 3-D printers are being used to create objects for transport, buildings, human organs, etc. There is a great deal of software used in 3-D printing. They can be from industrial grade to open source. The purpose of this book is to directly apply intellectual insights to use 3-D printing/additive manufacturing technology to solve our industrial problems.

My 30 plus years of experience across many different fields helps to provide unique perspectives on solving problems that we will be challenged with and help readers to gain an understanding of how this is done. I am a very motivated self-learner and am looking forward to sharing my past challenges with you, and I only hope that you are just as excited to read my book as I was when I started writing it.

3-D printing/additive manufacturing technology is of interest to me because it focuses on creating complex systems that cross multiple engineering disciplines. Every day as a chief engineer at The Boeing Company, GE, and COMAC, I work on systems from a wide array of disciplines that have been brought together to build enormously complex defense systems. This book will help to provide the skills that a Chief Engineer should possess and enable you to address manufacturing problems that you will be challenged by every single day.

By reading this book, you will be able to:

- Learn how some large companies have adopted 3-D printing/additive manufacturing technology for designing and producing complex parts for different industries.

- Determine all essential 3-D printing/additive manufacturing suppliers for supporting different industries.
- Learn how some companies have already converted their traditional design/production process to 3-D printing/additive manufacturing technology.
- Acquire the lifecycle strategy using 3-D printing/additive manufacturing.
- Identify all the influences of the cost drivers for using 3-D printing/additive manufacturing in different industries.

CHAPTER 1

Introduction

One of the most innovative inventions of the 21st century is three-dimensional (3-D) printing, sometimes referred to as additive manufacturing. A 3-D printer is a fascinating machine as it prints real-life three-dimensional objects instead of just printing some words or images on a piece of paper. 3-D printing is also referred to as additive manufacturing, making solid objects from a digital file. This object is printed using an additive process, which creates it by laying down layers of materials until the object is finally created.

A 3-D printer can produce complex shapes while using less material as compared to traditional manufacturing processes. 3-D printers are used to print any object irrespective of their material. Nowadays, through technical advancements, They can be used to create objects for transport, buildings, human organs, etc. There is a great deal of software used to print in three dimensions. They can be from the industrial grade to open source. The most commonly used software is Tinkercard, which is free and available online, enabling a 3-D object to be printed directly from the browser.

This book describes 3-D printing/additive manufacturing technology for 18 different industries, as follows:

(1) Chapter 2 3-D printing/additive manufacturing in the aerospace industry
(2) Chapter 3 3-D printing/additive manufacturing in the airplane parts industry
(3) Chapter 4 3-D printing/additive manufacturing in the automobile industry
(4) Chapter 5 3-D printing/additive manufacturing in the chemical industry
(5) Chapter 6 3-D printing/additive manufacturing in the construction industry
(6) Chapter 7 3-D printing/additive manufacturing in the dental care industry
(7) Chapter 8 3-D printing/additive manufacturing in the drone industry
(8) Chapter 9 3-D printing/additive manufacturing in the education industry

3D Printing
ISBN 978-0-323-99463-7
https://doi.org/10.1016/B978-0-323-99463-7.00001-3

(9) Chapter 10 3-D printing/additive manufacturing in the fashion industry
(10) Chapter 11 3-D printing/additive manufacturing in the food industry
(11) Chapter 12 3-D printing/additive manufacturing in the footwear industry
(12) Chapter 13 3-D printing/additive manufacturing in the healthcare industry
(13) Chapter 14 3-D printing/additive manufacturing in the hearing aid industry
(14) Chapter 15 3-D printing/additive manufacturing in the maritime industry
(15) Chapter 16 3-D printing/additive manufacturing in the mechanics industry
(16) Chapter 17 3-D printing/additive manufacturing in the movie industry
(17) Chapter 18 3-D printing/additive manufacturing in the tool and dye industry
(18) Chapter 19 3-D printing/additive manufacturing in the toy industry.

CHAPTER 2

3-D printing in the aerospace industry

Contents

2.1 What

A 3-D printer is a device that uses the method of layer-by-layer creation of a physical (solid) object using a digital 3-D model. 3-D printing can be carried out using various materials: plastic, metal, stem cells, and even food components. There are many 3-D printing technologies currently available, and new ones are constantly appearing. There are two main technologies for forming layers: laser and inkjet. The most commonly used are laser stereolithography and selective laser sintering.

3-D printing can be used in space in the following promising areas (Attaran, 2017):

1. 3-D printers for creating spare parts and tools on board the spaceship.

   The American space agency NASA and Made in Space sent the first 3-D printer to the International Space Station (ISS) in the fall of 2014 for the production of various parts, including: spare parts, instruments, and scientific equipment. The printer is able to make models layer by

3D Printing
ISBN 978-0-323-99463-7
https://doi.org/10.1016/B978-0-323-99463-7.00002-5

layer from polymers and other materials. 3-D models for creating objects are placed in the device's memory or transmitted from Earth if necessary.

This new technology is associated with grandiose prospects in optimizing work in space: from the simplest things, such as three-dimensional printing of broken parts, to the independent creation of robots, navigation systems, spacesuits, and research equipment.

2. 3-D printers for creating large-sized structures in space.

NASA, under the NIAC program, in 2013 allocated Tethers Unlimited, Inc. (TUI) $500,000 to further develop SpiderFab's automated assembly technology in space.

The technology is based on the Trusselator—a device that is a kind of cross between a 3-D printer and a knitting machine. The device is currently being successfully tested in the laboratory.

On one side of the cylindrical body there is a spool with thread (the device uses plastic as raw material, for example, carbon fiber), and on the other there is an extruder through which three main pipes of a future form or other structure are extruded. The truss is strengthened by winding it with a thread, and as a result, a robot about a meter long can create a truss tens of meters long.

A robot tracer using a manipulator and a special welding machine will be able to connect the original trusses into large complex structures and cover them with solar panels, reflective films, and perform other operations, depending on the objectives of the mission. There are different types of tracer, for example, it can produce round or square pipes of different diameters and thicknesses.

SpiderFab robots are equipped with an extruder that extrudes the finished plastic pipe with drum-containers of large capacity with raw materials, and manipulators for assembling the structure.

This technology makes it possible to manufacture, in space, very large, several kilometers long, spacecraft frames, antenna trusses, basic structures of solar power plants, huge telescopes, etc.

Currently, there is a huge surplus of structures being sent into space margin to withstand the overload at launch. Usually, in space, such heavy-duty structures are not needed, but a very large size is needed, for example, for telescopes and interferometers. SpiderFab devices allow such structures to be built: lightweight, large-sized, and with a low life cycle cost.

All necessary parts of the SpiderFab orbital production complex can be launched into space using existing launch vehicles. In fact, even with current technologies, SpiderFab allows for breakthrough projects, such as building space stations beyond the orbit of the Moon or solar power plants with a capacity of hundreds of megawatts. At the same time, the cost of structures produced using SpiderFab will be relatively small. One example of the use of SpiderFab is the construction of a space radio telescope worth $200 million with an antenna diameter of more than 100 m. Astronomers today can only dream of such an instrument, but SpiderFab technology could make this dream a reality in the coming decades.

3. 3-D printers for the construction of objects on other planets, for example on the Moon, including from improvised materials.

In 2011, NASA published its design for the construction of a lunar base with the participation of a large number of robots (excavators, bulldozers, shredders, etc.).

Now the European Space Agency has proposed an alternative project for 3-D printing the lunar base, using local soil as a building material.

The D-Shape printer manufactured by the British company Monolite is used for printing. On the Moon, the printer will be able to use the local soil, regolith, as a material.

Regolith is a loose, unevenly grained clastic-dusty layer several meters deep, consisting of fragments of igneous rocks, minerals, glass, and meteorites, that is well suited for construction.

Shown here is a 1.5-ton building block made by a D-Shape printer as a demonstration. The material used for printing is 99.8% similar to regolith, obtained from basalt rocks from a volcano in central Italy.

The Boeing 787 Dreamliner reportedly uses 30 3-D printed parts, a record in the industry. What's more, it was recently announced that General Electric is investing $ 50 million to 3-D print fuel nozzles for the next generation of LEAP jet engines. The main reason for the growing interest of the aerospace industry in 3-D printing technologies is the ability to produce significantly lighter parts. According to representatives of American Airlines, by reducing the weight of the aircraft by 500 g, more than 40,000 liters of fuel are saved annually.

(Sharke, P. (2017). How practical is 3-D metal printing? Mechanical Engineering, 139*(10), 44–49.)*

Vivek Saxena, vice president of aerospace at ICF International, a technical consulting firm, claims that 3-D printing technology currently accounts for approximately 0.0002% of global production. Specifically for the aerospace industry, 3-D printing has a 0.002% share of the $150 billion market. However, Saxena agrees with many experts that the market for 3-D-printed parts will reach $2 billion over the next decade. With these promising prospects, potential uses for 3-D printing look even more promising. The five most promising aerospace applications using these technologies are described below.

Considering that relatively small aircraft parts are already being 3-D printed, Boeing also considers it possible to create an entire wing using this method. At this stage in the development of 3-D printing, the production of large-sized objects is associated with a number of limitations, since with an increase in size, internal stress increases, which can cause distortion. However, the latest development from BAE Systems allows reinforced metal parts to be created by exposing each layer to ultrasonic waves. This helps to reduce stress, thereby making it possible in the future to 3-D print large objects such as aircraft wings.

In addition to 3-D printing fuel nozzles, General Electric is also actively developing the ability to 3-D print parts of the world's largest jet engine, the GE9X, for the next generation of Boeing 777X passenger aircraft. 3-D printing technologies can also prove useful in developing experimental prototypes for performance testing without having to spend large sums of money on numerical control equipment. A new joint project between Autodesk and Stratasys, in which a life-size turboprop engine was 3-D printed, showed how promising the use of 3-D printing is in the production of jet engine parts.

Currently, NASA's spacecraft contains about 70 3-D printed parts, but they are made on earth, which makes them very complicated supply chains and extended time frames. Manufacturing the required parts in space would significantly reduce costs and simplify planning, without having to send a rocket with the right parts and tools. 3-D printing of parts in space is currently being developed by research groups such as Made in Space and Lunar Buildings. Working with NASA, Made in Space is conducting zero-gravity experiments to test whether 3-D printing could be used on the ISS to produce the necessary parts and tools.

BAE Systems recently announced that 2040 of its new aircraft engineering concepts include equipment designed to create unmanned devices using a 3-D printer on board. The idea is that the aircraft arriving at the scene of the accident assesses the situation and reports the data to the control center. The information is then transmitted to on-board 3-D printers, which print out the unmanned aerial vehicles needed in this situation. These drones can monitor the situation or even conduct rescue operations. Although the concept is still in development, BAE Systems has already invested over $180 million in this research (Khan et al., 2020).

2.2 When

In his 1955 short story *A Necessary Thing*, Robert Shakesley described a configurator that two astronauts took with them on an interstellar expedition to print everything they might need in space, from ship parts to apple strudel for dessert. A little more than half a century has passed, and although reality has not surpassed this writer's imagination, it has come close to it.

NASA is developing a 3-D printer for printing spare parts directly on the ISS. Indeed, the development of 3-D printers in the near future could significantly affect the space industry in general and the prospects for the development of individual design bureaus in particular.

2.3 Where

NASA expects to use 3-D printing technology as a service to rapidly create preliminary prototypes. "3-D printing allows you to more accurately convey the concept of the mission. We can literally see what other developers are thinking," explains Tom Sonderstrom, head of IT at NASA's Jet Propulsion Laboratory. Using 3-D Platform As A Service (PaaS), engineers will be able to receive peer comments, design alternatives, and final prototype approval. In addition, this technology could help attract external specialists with original ideas, which will significantly reduce both the development time and costs.

Additive technologies, which seemed like science fiction a few decades ago, are now successfully used in a variety of fields. One of the most promising markets for their application is the aerospace industry, for which the problem of reducing production and time costs is especially acute. In addition, this industry is developing rapidly due to revolutionary scientific and technological achievements. The capabilities of 3-D technology are changing so massively that manufacturers will have to rethink their business models. New strategies are just emerging, but industry giants, such as NASA, SpaceX, Airbus, and Boeing, are already investing billions in 3-D printing prototypes, tooling, and engine parts for aircraft and spacecraft.

(From Khan, M.S. (2018). 3-D printing in transportation: already in action. TR News, *314.)*

2.4 Key benefits of additive technologies for aerospace manufacturing

- **Flexibility**: 3D printers are used to create components with complex geometries that cannot be manufactured using traditional manufacturing processes. We can confidently talk about the creation in the future of more and more perfect designs. Materials used in 3D printing (alloys, composites, polyamides, etc.) make it possible to manufacture unique products.
- Additive technologies are successfully solving a key problem for the industry—**reducing the weight of aircraft**. A 3D printer is the ideal tool for making topologically optimized components that increase payload and reduce fuel costs.
- Developing new and upgrading existing products takes less time and significantly **shortens the production cycle**.
- **Cost effective**: for highly specialized tasks, a 3-D printer allows for the use of fewer parts, materials, and accessories, while dramatically reducing the cost of small-scale components. With the development of technology, the cost of the finished product is falling, and the prices of printers have dropped by an order of magnitude over the past few years. According to forecasts, the cost of consumables will also decrease.
- Finally, additive technologies in some cases make it possible to **minimize waste** (compared to traditional processes) and improve the environmental friendliness and energy efficiency of aircraft.

2.5 Why

The possibilities of 3D printing will expand with the development of software. With the help of modern software, it will be possible not only to quickly create a complex model but also to automate the production process. Interactive models will make it possible to track and correct the slightest shortcomings in real time, calculate the safety and efficiency of all operations, and help in the understanding of how the equipment will behave in emergency situations. Thanks to advanced software technologies, suppliers, manufacturers, and designers will work in single information space and, in fact, communicate in a universal 3-D language. Today, most of the tests are carried out in real conditions, but in the coming decades, software will be able to solve increasing numbers of problems.

2.6 How

Additive technologies are an inevitable choice for a competitive enterprise. Today, most companies that implement them seek to reduce production costs and product development time without changing the supply chain and assortment. In the medium term, a strategy will come to the fore, involving the development of more complex, as well as new, highly functional products, which will lead to changes in the supply chain. Finally, in the long term, companies will be able to significantly improve manufacturability while eliminating middlemen from the supply chain. All of this will enable the use of radically new, more efficient business models.

2.7 Who

Astronauts in orbit often cannot provide themselves with everything they need and are forced to wait for the cargo that arrives at the ISS during scheduled flights. Unfortunately, during this time, the crew is not protected against accidents or breakdowns of important systems. Experiments with 3-D printing in space offer the potential to print the required parts if any should fail in space. This is very important for future flights to Mars and other planets, as, for a long time, the colonists will not be able to receive help from Earth. Therefore, in the upcoming expeditions, it is extremely important to use all available opportunities for the manufacture of products on board ships and space stations.

Such experiments have been conducted on the ISS since 2014. It was then that the Made in Space Zero G 3-D printer was delivered to the American segment of the station. The first printing took place on November 24, 2014, and marked a new era in the development of 3-D technology. The printed object was a part of the printer itself, the bezel of the print head, which symbolizes the ability to one day print a 3-D printer in space on a 3-D printer. In 2016, another Made in Space printer called the Additive Manufacturing Facility (AMF) was delivered to the ISS (Jun et al., 2017).

Since then, printing tests on the ISS have been taking place regularly. One of the latest ideas was the innovative plan of the Canadian company 3D4MD, which involved printing medical devices, such as splints or surgical instruments, on the ISS. To create devices such as individual splints for fractured fingers, 3D4MD developers could use, for example, measurements taken during the preparation of spacesuits and create a model on Earth. The 3D model can then be sent to the ISS, where the tire will be 3D printed (Lee et al., 2020).

Unable to conduct experiments in space, other powers organize experiments on earth. A development team in China recently successfully tested the first Zero Gravity 3D printer. Many difficult tests were carried out in the French city of Bordeaux.

2.8 3-D bioprinting in space

It is known that in open space there is electromagnetic and solar radiation, which has a destructive effect on biological tissues. In order for an astronaut to be able to endure all the hardships of their flight, the protection of the spacecraft alone is not sufficient—it is necessary to think about high-quality medical care, and if this does not help, then organs may need to be replaced.

That is why the American United Rocket and Space Corporation has agreed on an experiment on the use of a 3-D bioprinter on the ISS. It was developed by the United States laboratory 3-D Bioprinting Solutions, which specializes in bioprinting technologies. Scientists hope that a magnetic bioprinter will allow the creation of tissues and organs in space. The 3D bioprinter was completed and delivered to the ISS in 2018.

2.9 Construction of structures using 3-D printing

One of the most basic problems in the construction of buildings on extraterrestrial objects is the limited amount or lack of building material. The only available raw materials that do not need to be transported from Earth are local geological rocks. It is not surprising that scientists are solving the problem of using them in the construction of buildings.

For example, engineers at Northwestern University in the United States have found a way to use quality materials in situations where resources are limited. This is an additive manufacturing process from special materials that mimic the lunar and Martian regolith. These are strong and resilient materials that are produced using powdered substances that resemble rocks from the surfaces of the Moon and Mars.

It is likely that this technology will be used in the colonization of Mars. Due to extreme temperatures and high radiation on the surface of the "red planet," the first colonizers will need reliable shelters. NASA experts propose creating a "dome" of ice from the surface of Mars using 3-D printing, covered with a transparent membrane made of fluoroplastic-40. One of the main advantages of a water-based dwelling is that such walls provide protection from cosmic radiation but do not impede the penetration of light—this creates some comfort. In addition, other criteria were taken into account when choosing the most appropriate materials—their strength and reliability, and their ability to withstand the difficult conditions of Mars (Hoffman & Volpe, 2018).

The European Space Agency (ESA) is developing in the same direction and has already achieved some results. Scientists at the Austrian University of Applied Sciences in Wiener Neustadt managed to 3-D print a small needle and a corner wall made of JSC-Mars-1A material that mimics Martian soil.

2.10 Conclusion

In addition to serious research, there are also some quite unexpected projects. For example, experts from the ESA have discussed saving the souls of the first lunar colonists and planned to build a temple on the Moon using 3-D printing. This building, with the poetic name of the Temple of Eternal Light, will be located at the center of the residential complex for the first lunar settlers and will combine a prayer place and an observatory. The designers believe that the temples on the Moon will help revive the lost connection of mankind with space.

References

Attaran, M. (2017). The rise of 3-D printing: The advantages of additive manufacturing over traditional manufacturing. *Business Horizons, 60*(5), 677–688.

Hoffman, W., & Volpe, T. A. (2018). Internet of nuclear things: Managing the proliferation risks of 3-D printing technology. *Bulletin of the Atomic Scientists, 74*(2), 102–113.

Jun, S. Y., Sanz-Izquierdo, B., Parker, E. A., Bird, D., & McClelland, A. (2017). Manufacturing considerations in the 3-D printing of fractal antennas. *IEEE Transactions on Components, Packaging and Manufacturing Technology*, 7(11), 1891–1898.

Khan, M. S. (2018). *3-D printing in transportation: Already in action* (Vol. 314). TR News.

Khan, M. S., Sanchez, F., & Zhou, H. (2020). 3-D printing of concrete: Beyond horizons. *Cement and Concrete Research, 133*, 106070.

Lee, L., Burnett, A. M., Panos, J. G., Paudel, P., Keys, D., Ansari, H. M., & Yu, M. (2020). 3-D printed spectacles: Potential, challenges and the future. *Clinical and Experimental Optometry, 103*(5), 590–596.

Sharke, P. (2017). How practical is 3-D metal printing? *Mechanical Engineering, 139*(10), 44–49.

CHAPTER 3

3-D printing of airplane parts

Contents

3.1 3-D printing airplane parts

3-D printing is one of the most innovative inventions of the 21st century (Young & Sauter, 2020), which uses a printer, but not an ordinary printer, as it can print 3-D objects, not just text or images on a piece of paper. This innovative invention is also referred to as additive manufacturing, and creates solid objects from digital files. The object is printed using an additive process, which creates the object by laying down layers of material until the object is finally created (Baumers et al., 2017). Compared with traditional manufacturing processes, 3-D printers use less materials to produce complex shapes. 3-D printers can be used to print any object, regardless of the material. Today, through technological advancements, 3-D printers are used to manufacture objects such as vehicles, buildings, and human organs. In the 21st century, 3-D printers have become diversified, and their uses have expanded from commercial to household uses. The general public can buy 3-D printers and use them in their homes for personal use. They can print household items and even innovate new products that could be launched on the market in the future.

3-D printing or additive manufacturing is a development in industrial technology that uses transformative methods to make lighter and stronger

3D Printing
ISBN 978-0-323-99463-7
https://doi.org/10.1016/B978-0-323-99463-7.00003-7

parts and systems (Baumers et al., 2017). 3-D printing technologies provide multiple opportunities for different companies and involve multiple additive manufacturing technologies that can be used in the production process and have many benefits. As the transition from analog to digital processing made 3-D printing a success, this technological development has become a reality. A 3-D object scanner or CAD software (computer-aided design) is used to instruct the printer's hardware to deposit materials layer by layer in precise geometric shapes. Simply put, 3-D printing means adding materials until the desired object is created. On the contrary, when the object is produced by traditional methods, the excess material added to the object will be removed through different mechanical processes (such as milling, machining, engraving, etc.). Although additive manufacturing or 3-D printing has been on the market for decades, it remains new to many people. If the application of 3-D printing is delivered perfectly, it can ensure a perfect trio of smooth performance, ease of manufacturing, and complex geometric shapes. Thanks to this simplified procedure, many opportunities are waiting for those who can effectively embrace 3-D printing. This technology can be used for rapid prototyping, and all market areas that want to use cutting-edge technology could use this technology. Many multinational companies have started to adopt 3-D printers because they are more reliable and cost-effective, thus simplifying the process of manufacturing spare parts. There is a lot of software available for 3-D printing, which can be of an industrial grade or open source. The most commonly used software is Tinkercard, which is free, works online, and can print 3-D objects directly from the browser. The advantages of using a 3-D printer for spare parts are as follows.

3.2 Advantages

3-D printing has many benefits, no matter which industry it is used in. The primary advantage of 3-D printing is that it provides full customization, which means that consumers can print customized spare parts printers from 3-D according to their specifications. Parts can be manufactured effectively and economically using 3-D printing method. Customized 3-D printing allows manufacturers to produce spare parts that are no longer on the market. The second advantage is that 3-D printing technology avoids the stocking problem that many companies get into trouble due to stocking. 3-D printers allow parts to be produced on demand without the need for inventory. Digital printing has become the demand by many companies.

Spare parts production can be used for any type of products, whether it is home appliances, automobiles, airplanes, etc. 3-D printing parts allow the development of digital warehouse systems, in which products are provided digitally, and 3-D printing is used for printing technology, which can provide finished products in a short time. Many companies rely on physical inventory to meet the needs of the aftermarket, and digital inventory is their best solution. The advantage of 3-D printing is that it eliminates the business of traditional warehouses, which leads to the dematerialization of the supply chain. Another advantage of 3-D printing is that it simplifies the design and process of product manufacturing. No matter how complex the product design is, 3-D printers can make every task feasible by providing complete freedom in production and quantity. 3-D printing provides life to any type of complex project, and because objects can be printed in a single part, it can also reduce the assembly process. Moreover, additive manufacturing provides excellent finishing and high-quality finished spare parts.

3.3 Airplane parts

With advancements in technology, 3-D printers are being used commercially to manufacture aircraft parts. Many airports are innovating new concepts and installing printers at airports to make spare parts that are simple and affordable. This 3-D printing concept will help to control global warming and achieve zero carbon emissions, and it is producing many parts for commercial airliners such as the Boeing 777 or 787. These parts include aircraft shells, cockpit switches, circuit boards, luggage, seats, entertainment screens, windows, tray tables, turbofan duct mirror bosses, door lock shafts, etc. The prominent users of 3-D printing technology include the world's leading aircraft company, Airbus. Airbus' fleet has increased its variety of aircraft, built cost- and resource-saving aircraft, and achieved remarkable results: and the overall weight has been reduced by 45% utilizing 3-D printing technology. Excessive waste, saving 25% of production costs, reducing the number of parts from 10 to 1, reducing the weight of the aircraft by 4 kg and 16 door shafts (Vijayan, 2021). These additive manufacturing projects have brought various benefits to airlines, mainly because they reduce production and maintenance costs. 3-D printing is environmentally friendly and can also simplify logistics. If a printer is available at an airport, there will be fewer flight delays due to rapid manufacturing and replacement of parts. 3-D printers print mechanical and nonmechanical parts of airplanes, but the production of noncritical parts such as entertainment systems and tray tables is high because they more easily meet safety standards. These are ideal parts for 3-D printing because they can be produced quickly and on demand. 3-D printing technology has made a significant contribution to the aviation industry by producing lightweight structures and providing engines with complex geometric structures with clear aerodynamic characteristics. The core aspect of 3-D printing is that it can produce parts that can make the aircraft lightweight. Because seamless flight motion and control are based on weight, lightweight aircraft can improve the flight experience while reducing fuel consumption costs. According to statistics, if an aircraft loses 1 kilogram in weight, over its entire life cycle up to 25 tons of carbon dioxide emissions can be saved. There are many prominent examples of 3-D printed parts used in the aviation industry. In 2013, NASA printed and tested a rocket ejector during a fire test that produced 20,000 pounds of thrust (Nardozzo et al., 2019). In 2015, the FAA allowed commercial airliners to use printed parts for their jet engines (Singamneni et al., 2019). CFM's Feiyue engine consists of 19 3-D-printed fuel nozzles. In 2017, Boeing 787 used structural parts made of titanium wire produced by a 3-D printer.

Many airlines are studying other innovations or technologies that may affect future travel. They hope to introduce biological scanners to manage the physiological and nutritional needs of passengers. At the same time, they also hope to develop customized meal plans and supplements for passengers to eliminate passenger jet lag. Replacing manmade and machine-made products with 3-D-printed products could extend the life of an aircraft by 10—15 years while reducing maintenance and repair costs. Researchers aim to develop a printer that can use the right materials to produce flying parts and even an entire aircraft (Fig. 3.1). 3-D printing is an obvious choice to reduce the manufacturing process by setting structure and functional performance in a single supply chain process. Airlines using 3-D printers produce airplane parts as required, and will not produce any redundant parts or waste as often occurs in mass production. Many aircraft companies are advancing the innovations of the 3-D printing process, such as Emirates Aviation Engineering, Etihad Airways, BigRep, Singapore Airlines Engineering, etc.

Figure 3.1 3-D printing the entire body of an airplane.

3.4 What

A 3-D printer is a technologically advanced printer that prints objects as opposed to the conventional designs on a piece of paper by a regular printer. A 3-D printer can produce whatever object it has received through the instructions from a computer or through the Internet. Many companies have turned toward 3-D printing instead of using other materials to decrease their costs and increase efficiency. One of the breakthroughs in 3-D printing has been in producing aircraft parts related to both flying and nonflying aspects. Through this innovation, aircraft companies can lower their costs and carbon emissions, which benefits both the company and travelers. 3-D printing is a very cost-efficient process, and through various researches and innovations, some companies are trying to produce an entire airplane using 3-D printing technology.

3.5 Where

Many aircraft and engineering companies have 3-D printing technology to produce aircraft products, and are conducting new research to install portable printers at each airport to simplify the production process. 3-D printing companies, including SIA Engineering, work with Stratasys, which makes 3-D parts for aircraft. The Boeing Company is a leading aircraft manufacturer with a division in California that produces parts through 3-D technology. Emirates Aviation Engineering is the leading airline in the UAE, and works with 3-D systems to produce video monitor shields for its aircraft through the use of 3-D printing technology. Many other airlines have followed suit.

3.6 Who

In most cases, airlines have developed and installed 3-D printing technology in their operations to meet the requirements of easy production and replacement of parts, while contributing lower production costs and more profits to the company. By contrast, some airlines have taken the risk and joined 3-D companies and are producing spare parts. The UAE's leading airline, Emirates, has joined forces with 3-D systems to produce parts. Boeing has established 3-D printing technology in its own company to reduce its production costs. BigRep is a company that produces 3-D printers, and many airlines have cooperated with them to provide 3-D products, including Airbus and others.

3.7 Why

3-D printing technology is one of the most important innovations this century, as it does not waste products after production, reduces production costs, and reduces carbon emissions. Therefore, most airlines are turning to 3-D printing to reduce costs and improve the in-flight experience.

3.8 When

The first 3-D printed product was manufactured and tested on a commercial jet airliner in 2014. A metallic titanium bracket was tested on an Airbus aircraft, and it is well known that this product has achieved success and still fulfills its requirements. After this successful test, other airlines began to produce and use 3-D printed products on their commercial jet airliners (Fig. 3.2).

3.9 How

After having to deal with higher fuel consumption and increasing costs, many airlines began to study efficient production methods. After many years, they finally started to produce 3-D-printed spare parts for aircraft,

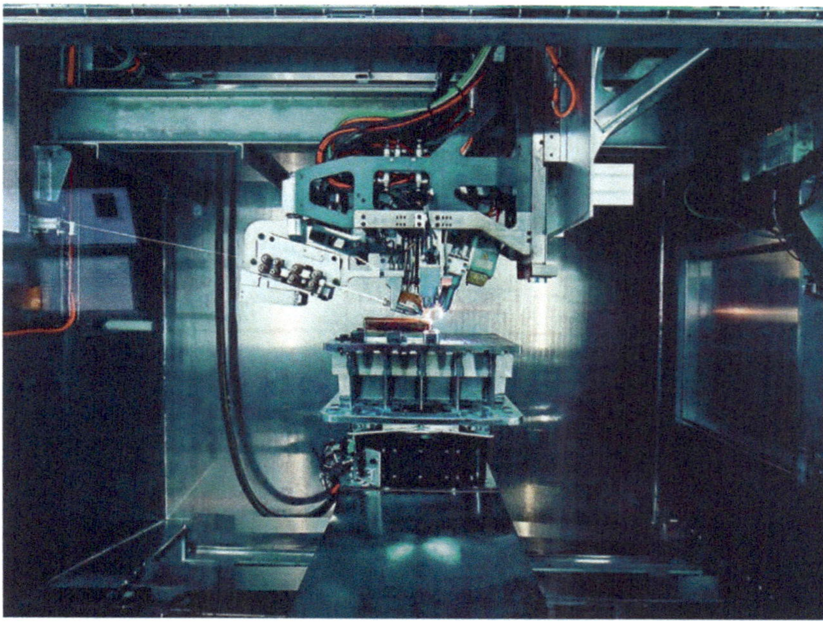

Figure 3.2 3-D printing of titanium parts for Boeing 787 Dreamliner.

which reduced fuel consumption and improved the passenger's flying experience. In 2014, the first 3-D product was successfully tested on the Airbus 320, and other companies followed suit (Joshi & Sheikh, 2015).

References

Baumers, M., Duflou, J. R., Flanagan, W., Gutowski, T. G., Kellens, K., & Lifset, R. (2017). Charting the environmental dimensions of additive manufacturing and 3D printing. *Journal of Industrial Ecology, 21*(S1), S9–S14.

Joshi, S. C., & Sheikh, A. A. (2015). 3D printing in aerospace and its long-term sustainability. *Virtual and Physical Prototyping, 10*(4), 175–185.

Nardozzo, P. K., Popkin, S., & Smith, J. R. (2019). Feasibility study for hypersonic flight test using a reusable hybrid rocket motor. In *AIAA propulsion and energy 2019 forum* (p. 3838).

Singamneni, S., Yifan, L. V., Hewitt, A., Chalk, R., Thomas, W., & Jordison, D. (2019). Additive manufacturing for the aircraft industry: A review. *The Journal of Aeronautics and Aerospace Engineering, 8*(1).

Vijayan, V. (2021). *3D printing in aviation.* EOS GmbH. Eos.info. Retrieved 5 April 2021, from https://www.eos.info/en/3d-printing-examples-applications/aerospace-3d-printing/airbus-case-study.

Young, A., & Sauter, M. (2020). *3D printing, E-cigarettes among the most important inventions of the 21st century.* Retrieved 5 April 2021, from: Usatoday.com https://www.usatoday.com/story/money/2020/01/09/21-most-important-inventions-of-the-21st-century/40934825/.

Further reading

Marks, P. (2011). *3D printing takes off with the world's first printed plane.*

Savastano, M., Amendola, C., Fabrizio, D., & Massaroni, E. (2016). 3-D printing in the spare parts supply chain: An explorative study in the automotive industry. In *Digitally supported innovation* (pp. 153–170). Cham: Springer.

CHAPTER 4

3-D printing in the auto industry

Contents

4.1 When: since when has 3-D printing been used in the auto industry

3-D printing has been used in the automotive industry for 30 years to produce some parts, especially in the prototyping and verification phases of projects, and to create demonstration projects and working parts that are difficult to print at a low price for vehicle and engine testing. Over the past 2 years, 3-D printing has become an absolute necessity for the development of new models. All major car manufacturers have strengthened and expanded its use, including making molds and special tools, and they are now beginning to use it in their manufacturing processes. 3-D printing is widely used in racing cars, where it allows for lighter parts.

3-D printers will increasingly serve industrial production because they have the advantages of simplifying the assembly process, and reducing production time and costs. Another unquestionable advantage is the possibility of creating new high-performance materials, experimenting with combinations of elements with different characteristics, such as titanium, aluminum, nickel, and steel. Compared with traditional manufacturing methods, metal 3-D printing requires less energy and materials to be used in the manufacture of components. In addition, 3-D printing can significantly shorten the time to market (Shahrubudin et al., 2019).

3D Printing
ISBN 978-0-323-99463-7
https://doi.org/10.1016/B978-0-323-99463-7.00004-9

In Italy, Ferrari, Lamborghini, Ducati, and Agusta were the first to adopt 3-D printing technology, using in-house professional printers such as EOS and Stratasys, as well as rapid prototyping and additive manufacturing services provided by companies such as CPC Group, CRP Group, and Energy Group. HSL, Skorpion, Proto Service, Zare, and other companies print metal and plastic parts. Lamborghini uses 3-D printers to produce scale models and functional prototypes, reducing the production time and cost of its cars, and expanding its product range. To celebrate the 50th anniversary of the brand, 3-D printing has made it possible to produce just nine Venenos, priced at three million euros. Ferrari also sold a total of only 10 F60 America models to celebrate the 60th anniversary of its first subsidiary in the United States (Attaran, 2017).

4.2 How: how has 3-D printing impacted the racing world?

3-D printing has also significantly entered the world of car racing. As early as 2012, the Australian V8 supercar team of Nissan Motorsports used 3-D printing in cooperation with Evok3D to make major changes to the cooling system and aerodynamics within a few weeks, and it is now used by most Formula One teams. In addition to luxury cars and sports cars, 3-D printing is also used to produce custom parts for mass-produced cars, such as shift knobs, dashboards, etc. For many years, all major 3-D printing processes using polymers and metals have been used in the automotive industry to produce prototypes of parts, engines, and car bodies. The most recent innovation is that now that additive manufacturing has reduced costs and reached production speeds, and even small parts, components, and accessories for end-use applications can be implemented more frequently. For example, Fiat Chrysler (FCA) also uses 3-D printing to make transparent plastic models of lubrication systems to check the oil flow in engines, and is the main customer of 3D Systems Italia's 3-D printing services in Pinerolo. Since the launch of the Ford GT Heritage Edition in 2006 to celebrate its 100th anniversary, Ford has been using 3-D printing in its additive manufacturing center. Ford produces up to 100,000 3-D-printed components per year and creates final prototype components (Savastano et al., 2016).

Another important market is spare parts with a low turnover, or vehicles that are no longer in production, especially those of small size that are quite complex. For example, the Italian startup Musp Garages uses 3-D printing to produce mechanical parts for vintage cars that are no longer on the market, and provides spare parts for additive manufacturing related to the microcasting process. The consulting firm Carlisle predicts that 30% of spare parts will be printed on demand in the next few years. 3-D printing will greatly help to simplify the management of spare parts and significantly reduce the size of spare parts warehouses. Cars whose bodies are completely manufactured by 3-D printing have also begun to appear. The first was the 2011 Canadian concept car Urbee, and the second was the Local Motors 3D Strati, which was printed in only 44 h during the Chicago International Manufacturing Technology Show in September 2014. San Francisco start-up company Divergent Microfactories printed a 3-D Blade, a 700-horsepower methane and gasoline-powered supercar that weighs only 635 kg, and it is also part of a larger project that aims to fundamentally reduce the use of polluting resources generated during the manufacturing process and make the production of cars in microfactories economically viable. The characteristic element of Blade is the proprietary Node technology: a 3-D printed aluminum structure that allows one to connect the carbon fiber elements that make up the frame. Divergent Microfactories will sell a small number of products, which will be built in its microfactories. The Chinese company Sanya Sihai 3D also displayed a 3-D printed electric minicar Shuya, which can reach a speed of 40 km per hour and weighs 500 kg, priced at about 1700 US dollars.

4.3 What: what are the benefits of 3-D materials and technologies?

Innovative desktop 3-D materials and technology enable engineering and design teams to significantly shorten the distance between the idea and the final product. Prototypes of individual internal elements such as dashboards can be printed directly on site within a few hours to test their applicability and make adjustments immediately if necessary. The new special synthetic resin used for prototyping allows us to reduce production time by more than 70%. Therefore, the product development process can be speeded up the iteration cycle and can be strengthened also, with the elimination of dependence on external suppliers, long delivery times, and related generally high production and storage costs. Implementing adjustments immediately and testing overhauled parts almost on the same day provides great potential for the automotive industry, and rapid verification is essential. Closing a car production line for a short period of time can cost thousands or even tens of thousands of euros. As a result, shutting down spare parts deliveries during production could cost automakers 100 million euros per week.

In addition to faster and cheaper manufacturing and customization processes, 3-D printing also provides new options for low-volume shapes and designs, and customization. For example, 3-D printing can produce customized parts in small batches at an affordable cost without the need to customize the entire production line or a large amount of labor. Especially for smaller workshops, eliminating high initial investment opens up entirely new possibilities.

Not only is the use of new and innovative 3-D printing technology opening up a variety of new possibilities for the automotive industry, it is also combined with traditional craftsmanship. An example is the new version of the famous Volkswagen minibus from 1962. The unusual interweaving shape of the aluminum rims of the new version of the Volkswagen minibus requires a specially designed rim combination. The hub caps were first 3-D printed using an Sterolithography (SLA) process in cooperation with Formlabs, and then galvanized to obtain the look and feel of real metal parts to achieve the appropriate design (Kothman & Faber, 2016).

The combination of polymer-printed parts and processes (such as lost wax casting, sand casting, or direct low-temperature casting) to produce specific metal parts with unique surface finishes provides manufacturers with a cheaper and more detailed alternative, and it is not directly used for metal and other traditional 3-D printing, that are usually extremely expensive manufacturing processes. Combining 3-D printing with other processes can take advantage of innovative 3-D printing technologies with polymers (such as time, cost, ease of use, precision, and rapid iteration) to create other more complex materials, such as metals.

Through the electroplating process, one can combine 3-D-printed plastic parts with the excellent mechanical and chemical properties of metal. In this process, a thin layer of the required metal is electrochemically deposited on the surface of the 3-D printed part. The result is a real metal surface. In addition to giving a metallic appearance, electroplating also significantly enhances the strength of the underlying components, and improves the wear resistance, UV resistance, and corrosion resistance of the material. For example, the tensile strength of SLA 3-D-printed parts has increased by an average of three times, the bending strength has increased by nearly 10 times, and the stiffness has increased by four times (Khosravani & Reinicke, 2020).

(From Source: https://blog.mahindrafirstchoice.com/important-car-parts-and-how-they-work/).

4.4 Why: why should automotive companies use 3-D technology?

Using 3-D printing technology, the entire dashboard, external grille, and nose bridge can be printed for prototypes to be tested on roads and in wind tunnels. It is also possible to print a prototype as a master to produce a silicone mold, allowing colored plastic to be injected into it to obtain a pre-series (20/30 pieces).

The main reasons for 3-D technology being so relevant to the automotive industry are as follows:

- Compared with many other traditional technologies, the use of 3-D printers can create prototypes in a short time and at a competitive cost, thus accelerating the product development process.
- Through 3-D printing, waste can be minimized. The materials used are only those necessary for the manufacture of automotive parts. This technology has a huge cost advantage in metal material processing. All the extra costs of wasting materials are avoided.
- 3-D printing in the automotive sector makes many customization requirements possible. The creation of internal components or some external functions does not require enhanced costs, even in individual customization.
- The use of 3-D technology facilitates the replacement of worn parts. By combining 3-D scanning and additive manufacturing, rare spare parts can be optimized and replicated.

Analysts believe that 3-D printing cars is used to promote a technological development concept and drive consumption through this concept, rather than using 3-D printing to produce cars on a larger scale. However, as an emerging technology, 3-D printing has its unique advantages, and there are still many successful application cases for the automotive industry. The following analyzes the 3-D printing technology of the automobile industry from the different stages of automobile design, manufacturing, and consumption. In the automotive design stage, 3-D printing can quickly produce conceptual models and functional prototypes, significantly shortening the verification time of auto parts in the initial stage of product development, thereby shortening the automotive development cycle and minimizing cost and design waste. For example, Lamborghini uses 3-D printing technology to develop engine pipelines to complete complex structures in a short time. In the automobile manufacturing stage, the exterior of the car including lights, bumpers, signs, fenders, etc., and the interior of the car, including the audio system, steering wheel, instrument panel, joystick, etc., as well as the chassis and body, have successfully applied 3-D printing technology. According to reports, the application of 3-D printing technology has played an important role in reducing the production cycle and production costs.

4.5 Where: where is 3-D printing being used?

Many auto manufacturing companies are making full use of 3-D printing technology to manufacture auto parts. In addition to auto parts, this technology is also used in tools. One of the first companies to adopt 3-D technology was Ford Motor Company. The latest 2020 Shelby GT500 makes extensive use of 3-D printing technology. Engineers ensure that the best downforce, braking, and cooling goals are achieved using parts created with 3-D technology. Another major example is the Volkswagen Autoeuropa, whose jigs and fixtures are entirely made using 3-D printing technology.

Some people have imagined that the development of electric vehicles has replaced traditional engines, and the entire vehicle structure has undergone major changes. The motor can be installed in the hub and connected to the battery with a wire. There is no need for traditional engine systems and transmission systems. Instead, vehicle communication, audio and video systems, etc. can be used. Not only through the use of lightweight materials, but also through the design of complex components, can lightweight vehicles be achieved. In summary, the advantages of 3-D printing technology will show their possibilities in the future automobile manufacturing industry, mainly in the following four aspects.

The supply of 3-D-printed parts is still one of the important development directions of the automotive industry for the future. Unlike in the past, as 3-D printing technology matures, it will provide increasingly diverse types of parts. In the future, in the entire parts supply chain, especially for models that have been discontinued, there may no longer be physical parts storage. As long as the relevant data for each part are kept, it will be able to be printed out when needed. Although 3-D printing technology has not been developed for a long time, its speed is impressive, and its application in the automobile manufacturing industry is becoming increasingly extensive (Khosravani & Reinicke, 2020).

(From Source: https://www.thepostcity.com/3-D-printing-in-automobile-manufacturing-the-futuristic-approach-towards-growth/).

4.6 Who: who is getting the benefit from 3-D technology?

It is undeniable that 3-D printing technology has great potential in the automobile manufacturing industry and has advantages that traditional manufacturing methods cannot match. For example, the sheet metal covering parts of the car body need to go through multiple complicated processes such as die stamping, three-dimensional cutting, trimming, punching, etc. However, if 3-D printing technology is used, the mold design and manufacturing can be designed on CAD to simplify the process and manufacturing cycle. However, 3-D printing technology also has many limitations. At present, 3-D printing cannot be compared with traditional industrialized mass production, and cannot achieve large-scale standardized operations. For the automobile manufacturing industry that pursues high-efficiency production, the two are undoubtedly incompatible at this point. In addition, 3-D printing technology also has limitations in materials. Expensive and rare materials cannot be printed, and the lack of compound talents in this area makes 3-D printing technology temporarily unable to be widely used in automobile manufacturing. In the context of global energy conservation, emission reduction, and environmental protection, major auto companies have responded to laws and regulations formulated by various countries to improve engine thermal efficiency and reduce carbon dioxide emissions. In addition, the vehicles have become lighter. As mentioned earlier, the weight of the crankshaft manufactured by Honda using 3-D printing technology can be reduced by 30%. It is assumed that the weight of all vehicle components could be significantly reduced. In this case, fuel consumption also could be reduced to a large extent, and the goals of reducing emissions and increasing cruising range can be achieved (Khosravani & Reinicke, 2020). However, after so many years, 3-D printing technology has not fully replaced traditional automobile manufacturing. In concept advocacy can be greater than practical application. In the traditional automobile manufacturing industry, 3-D printing technology has been successfully applied to the manufacture of common parts such as signs, fenders, steering wheels, dashboards, etc. However, 3-D printing technology for high-performance parts remains to be explored. However, in March of 2021, Honda's R&D department and software company Autodesk launched a cooperation in which both parties designed the crankshaft and used 3-D printing technology to make samples. It has been reported that compared with traditional crankshafts, this crankshaft can achieve a significant weight reduction of 30%. However, the crankshaft, which is the core component of the engine, has high working strength and the degree of wear resistance needs to be further tested. If this crankshaft can be successfully applied to mass-produced cars, does it mean that increasingly high-performance parts can be manufactured with 3-D printing technology? The answer is yes, but this is currently not realistic (Khosravani & Reinicke, 2020). It is undeniable that 3-D printing technology has great potential in the automobile manufacturing industry and that it has advantages over traditional manufacturing methods. For example, the sheet metal covering parts of the car body need to go through multiple complex processes such as die stamping, three-dimensional cutting, trimming, punching, etc. However, if 3-D printing technology is used, the mold design and manufacturing can be designed on CAD to simplify the process and manufacturing cycle. However, 3-D printing technology has many limitations. At present, 3-D printing cannot compete with traditional industrialized mass production, and it cannot achieve large-scale standardized operations. For the automobile manufacturing industry that pursues high-efficiency production, the two are undoubtedly incompatible on this point. In addition, 3-D printing technology also has limitations in materials. Expensive and rare materials cannot be printed, and the lack of compound talents in this field makes 3-D printing technology temporarily unable to be widely used in automobile manufacturing (Khosravani & Reinicke, 2020).

References

Attaran, M. (2017). The rise of 3-D printing: The advantages of additive manufacturing over traditional manufacturing. *Business Horizons, 60*(5), 677–688.

Khosravani, M. R., & Reinicke, T. (2020). On the environmental impacts of 3D printing technology. *Applied Materials Today, 20*, 100689.

Kothman, I., & Faber, N. (2016). How 3D printing technology changes the rules of the game. *Journal of Manufacturing Technology Management, 27*(7), 932–943.

Savastano, M., Amendola, C., Fabrizio, D., & Massaroni, E. (2016). 3-D printing in the spare parts supply chain: An explorative study in the automotive industry. In *Digitally supported innovation* (pp. 153–170). Cham: Springer.

Shahrubudin, N., Lee, T. C., & Ramlan, R. (2019). An overview of 3D printing technology: Technological, materials, and applications. *Procedia Manufacturing, 35*, 1286–1296.

Further reading

4 ways 3D printing is transforming the maritime industry.(2017). Retrieved 26 February 2021, from https://www.linkedin.com/pulse/4-ways-3d-printing-transforming-maritime-industry-chuck-fox.

CHAPTER 5

3-D printing in the chemical industry

Contents

5.1 Introduction

The field of chemistry has undergone tremendous development. As published in the journal *Science*, three-dimensional (3-D) printers that can synthesize molecules and make compounds in just a few hours have been designed. If done by traditional methods, this would take years. The traditional supply chain has been disrupted, and where demand increases, 3-D printing has proven to be almost the only option to quickly repair loopholes in the production of necessary products, even taking into account that the requirements for medical products are very strict. 3-D printing is the most eye-catching and direct embodiment of additive technology. It is a method of stacking materials layer by layer in the production of metal or polymer parts.

3D Printing
ISBN 978-0-323-99463-7
https://doi.org/10.1016/B978-0-323-99463-7.00005-0

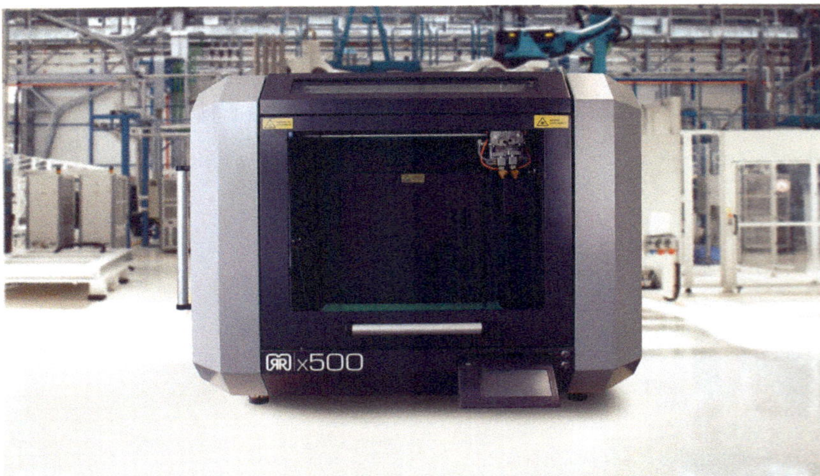

(Hanaphy, P., Hanaphy, P., Sertoglu, K., Sertoglu, K., Sertoglu, K., &Sertoglu, K. et al. (2021). German RepRap, DuPont, dow chemical company and 3DChimera demonstrate the possibilities of 3D printing at RAPID + TCT 2019 - 3D Printing Industry. Retrieved 27 March 2021, from: https://3dprintingindustry.com/news/german-reprap-dupont-dow-chemical-company-and-3dchimera-demonstrate-the-possibilities-of-3d-printing-at-rapid-tct-2019-155929/).

5.2 What

Although the prospects for wide application of 3-D printing in industry, medicine, and construction are optimistic, the use of this technology is mainly related to the development of product prototypes, as well as simplifying the manufacturing process and obtaining single or small-scale products.

When can 3-D printing be used in chemical laboratories and trial production? The best option is when multiple parts with complex geometries need to be obtained, and when it is not certain whether this geometry will be retained in the future after testing. In these instances it makes more sense to use 3-D printing instead of casting.

Of course, casting into silicone or metal molds is an excellent manufacturing method, and parts produced by this method are usually stronger than parts obtained by 3-D printing. However, when strength characteristics are not as important as saving time or money in formulating any structure, 3-D printing is preferable.

The use of 3-D printers in mass production has raised questions because of the relatively high cost of such devices (especially for high-precision printing or metal printing), low productivity, and the inability to combine different materials until now. Organizing such a business requires a lot of investment, although training experts is not a particularly difficult task.

In small-scale production and casting model and mold manufacturing, the use of 3-D technology is reasonable. Products made of transparent materials allow clearly observation of the processes happening inside the equipment. An important area of 3-D printing is the rapid production of complex parts (for example, line printing of spacecraft parts, which can save tens or even hundreds of standard hours in the production of key parts) (Konarova et al., 2017).

5.3 When

The first predecessor of 3-D printers can be traced back to the early 1980s, when Dr. Hideo Kodama of Japan developed a rapid prototyping system using photopolymers. The working principle of this device is the same as that of modern 3-D printers—the printed objects are created layer by layer according to the model. The real breakthrough in this field is the stereolithography technology invented by Charles Hull in 1984. With this technology, objects can be produced based on digital projections (images) on a 3-D printer. The material is also used as a photopolymer, a liquid substance based on acrylic acid. Under the influence of an ultraviolet laser beam, the material immediately solidifies and becomes a plastic object, forming the desired shape. This invention has revolutionized the development community, creating prototypes at a significantly lower cost.

5.4 Who

In the early 1990s, Charles Hull's company 3-D Systems Corporation (still one of the industry leaders) began using photopolymer laser stereolithography (SLA) technology to produce equipment. At the same time, the DTM startup obtained a patent for the selective laser sintering (SLS) technology invented by Karl Deckard (University of Texas), which exposes powder rather than liquid materials to laser light. The third technology, fused deposition modeling (FDM), was developed during this period under the direction of Scott Crump of Stratasys and is still used primarily for basic 3-D printers today. Of course, the original device was not perfect. As the material solidifies, objects often bend, despite the drawbacks and high cost of these techniques.

Over the next decade, several important events occurred in the application of 3-D printing technology in medicine. The first of these was the successful implantation of a 3-D printed bladder. The finished organ was covered by the patient's cells, which reduced the possibility of rejection of the implant to zero. Subsequently, it was announced that a fully functional small kidney copy had been created, but this claim was later refuted. Even now, scientists admit that it is still a few years or even decades before it will be possible to create a fully functional and complex organ suitable for transplantation, such as the kidney or liver. Various researchers have developed kidney models, but the main problem remains the large and small vascular system, because without it, the organ will not be able to function in the patient's body. However, significant progress has been made in 3-D printing of simpler organs, cartilage, and blood vessels using only human cells as materials, in addition to bone implants (An et al., 2019). With the introduction of living tissue printing and the development of complex prostheses that do not require assembly, this research line is developing rapidly.

(From Chabaud, C. (2021). Chemical Industry in connestion with 3D printing. Retrieved 27 March 2021, from https://www.sculpteo.com/blog/2015/10/28/3d-printing-for-the-chemical-industry/).

5.5 Where

In 2019, the global 3-D printing market was estimated to be USD 13.7 billion; by 2025, it is expected to increase to USD 63.46 billion. The annual growth rate will be around 30%. The United States is developing unique technologies in this field. For example, a method for printing large heavy metal products has been developed, with no analogues in the world (Schiffer & Manthiram, 2017).

The main features of 3-D printing in part design and the limitations of the possibility of using this technology are as follows:

1. The thickness of the layer formed by the working tool (head) of the 3-D printer in one pass;
2. The thickness of the material that can be manufactured (for example, limited by the extruder nozzle diameter, Electron Beam Freeform Fabrication (EBF) wire diameter, etc.);
3. The characteristics of the material obtained through a specific type of printing;
4. The time it takes to produce a part;
5. The size of the working area of the 3-D printer.
6. When using chemical technology, harsh environments, high pressures and high temperatures, often have to be dealt with, which greatly limits the use of polymers used in 3-D printing by most printers (Lin et al., 2018).

5.6 How

ABS and PLA are the most commonly used 3-D wires as materials for 3-D printers. However, in some cases, other types of filaments must be used, because the previous filaments lacked sufficient mechanical, physical, or chemical properties to meet more demanding applications. Advanced materials are 3-D filaments with mechanical properties and heat resistance or contact with chemical elements, which makes them superior to other 3-D materials. The types of 3-D filaments described in this chapter are common in industrial, aerospace, and/or automotive 3-D printing, where the manufactured parts or components need to go through a very demanding manufacturing process. BASF Filemento Ultrafuse is one of the most advanced and innovative materials markets. The German chemical industry giant BASF successfully placed a filament containing 90% steel on the market. This is a kind of filament made of steel powder and plastic binder material, which can be used for printing. Once printed, these parts undergo a sintering process in which the plastic binder disappears and metal bonding of the steel is achieved through sintering. It is an ideal filament for manufacturing spare automobile parts or machine mechanical parts through 3-D printer type FDM, without the need to invest in expensive EBM or SLS metal machines (Parra-Cabrera et al., 2018). The 3-D industry continues to grow. The revolution that created this printing press was accompanied by constant innovations in printed materials. Acrylonitrile Butadiene Styrene (ABS) carbon fiber is a kind of injection molded parts widely used in the automotive industry. By reinforcing this 3-D filament with carbon fiber, it is possible to increase its rigidity and tensile strength, while reducing the weight and flammability. This makes it an ideal choice for parts that require a lot of effort, such as mudguards and support elements in the automotive industry, liquid containers, and other components with high mechanical requirements. Carbon fiber-filled Poly Ethylene Terephalate Glycol (PETG) is a nontoxic thermoplastic polymer widely used in the textile industry and packaging manufacturing. By adding carbon fiber, extremely durable materials that are very resistant to high mechanical stress are created. Because of these characteristics, it is almost at the same level as other more advanced materials, which is why it has become a very popular material to replace metal parts subjected to moderate mechanical stress. It also performs well in severe weather conditions and temperatures, and responds well to chemical reagents (Parra-Cabrera et al., 2018). Carbon fiber-reinforced nylon is a material for 3-D printers, which is very suitable for manufacturing parts subject to heavy mechanical stress. Compared with other parts made of fiber-reinforced filaments, it has higher durability. It also has good properties such as good adhesion and low deformation after printing.

5.6.1 Polyoxymethylene (POM) 3-D filament

POM, also known as acetal or polyacetal, is a material with appropriate stiffness, low friction, high impact resistance. It is a material with high elasticity, heat resistance and high chemical resistance. Due to its intense conflict, POM is very commonly used in gear systems, ball bearings and other parts where fluidity inside the assembly is critical.

5.6.2 3-D Poly Ether Ether Ketone (PEEK filament

PEEK is an advanced printing material with excellent mechanical properties and high temperature resistance. It is used in very demanding applications and environments, such as fuel processing plants or component manufacturing in the aerospace or automotive industries. It has an excellent response to chemical reagents. In addition, it is also suitable for food applications and even medical applications by allowing sterilization.

5.6.3 Limitations of 3-D applications in chemistry

So far, 3-D printing technology has many obvious limitations and shortcomings, and it cannot be widely used in high-precision chemical work.

It is hoped that these difficulties are temporary. The main problem lies in the technical capabilities of a single device.

For some types of additive technologies, the layer thickness or the thickness of the cured material bundle can cause significant errors in the shape of the part, because the formation of layer "steps" in any area with a cylindrical or spherical surface is impossible to avoid. Such parts need to be processed by traditional methods (turning, milling), as it has been proven that neither time nor labor costs can be saved when manufacturing complex geometric products. So, what is the point of this expensive technology?

Our Engineering Center is open to innovation; we always try to test new technologies that can help us solve current problems and optimize processes. Unfortunately, we were unable to find a 3D printer in DuPont that can print scales: the nozzle holes are 2 mm in diameter and require no further processing. It is difficult to ensure acceptable surface roughness because the material deposits in layers, sinters and flows unevenly during the melting process.

The part is obtained by using SLS technology (powder sintering layer by layer), then the powder residue is removed from all holes and the outer cylindrical surface is machined on a lathe, because a smooth cylindrical surface cannot be formed. It turns out that parts made in this way are very strong and airtight (PA 2200 polyamide is used). Finally, for creating a small series of parts (25–50 items), the 3-D printing method has proved to be the most acceptable in terms of price–quality–time ratio.

It takes more than 2 months to manufacture a single injection mold for this product, and the cost exceeds one million rubles (1Russian Ruble = 0.012 US$), which is acceptable and reasonable in the production of 2000–3000 products per month, but not for a single or few items.

5.6.4 Specific shortcomings of 3-D technology in chemistry

Another problem with 3-D printing is the need for support in this type of printing. No solid bulk or liquid materials (SLS/SLA technology) are solidified layer by layer. This is the additional supporting technical part of the material created by the printer to form the part of the material application layer that is not connected to the main feature. The printing of the next layer must be connected to it. In addition, a "support" is formed so that the wall and the printing table are at an angle.

These support elements must be removed by additional mechanical treatment and uncured powder using SLS technology or uncured photopolymer using SLA technology. For example, certain materials (such as ABS plastic) can be treated with the vapor of a low-boiling solvent (such as acetone) to make the surface smooth and rounded. By creating an integral shell of the part, the risk of delamination is reduced. Naturally, this is due to the partial dissolution of the surface layer, which destroys the geometry of the location. As an example, we will show a detail that the author wishes to create but has so far been unable to manufacture using additive technology. Honeycomb carrier is used for catalyst thin-walled parts (wall thickness: 0.05–0.1 mm; material: pure AD0 aluminum (or similar); transparent cell size: ~ 2 × 3.5 mm). The supported catalyst is a critical part of the final product, it must have sufficient thermal resistance, mechanical properties and vibration resistance.

Companies using aluminum powder 3-D printers have stated that their wall thickness is limited to 0.3 mm; the minimum wall thickness required is 0.8–1.0 mm. The use of this thickness in our design is unacceptable because it will result in a significant decrease in the quality of the final product and increase the aerodynamic resistance to airflow.

References

An, Y. J., Guo, C. F., Zhang, M., & Zhong, Z. P. (2019). Investigation on characteristics of 3D printing using Nostocsphaeroides biomass. *Journal of the Science of Food and Agriculture, 99*(2), 639–646.

Chabaud, C. (2021). *Chemical Industry in connestion with 3D printing*. Retrieved 27 March 2021, from https://www.sculpteo.com/blog/2015/10/28/3d-printing-for-the-chemical-industry/.

Hanaphy, P., Hanaphy, P., Sertoglu, K., Sertoglu, K., Sertoglu, K., & Sertoglu, K., et al. (2021). *German RepRap, DuPont, dow chemical company and 3DChimera demonstrate the possibilities of 3D printing at RAPID + TCT 2019*. 3D Printing Industry. Retrieved 27 March 2021, from: https://3dprintingindustry.com/news/german-reprap-dupont-dow-chemical-company-and-3dchimera-demonstrate-the-possibilities-of-3d-printing-at-rapid-tct-2019-155929/.

Konarova, M., Aslam, W., Ge, L., Ma, Q., Tang, F., Rudolph, V., & Beltramini, J. N. (2017). Enabling process intensification by 3D printing of catalytic structures. *ChemCatChem, 9*(21), 4132–4138.

Lin, J. C., Wu, X., Yang, W., Zhao, R. X., & Qiao, L. G. (2018). Application of PO and R-SAC mortar for 3D printing in construction. *IOP Conference Series: Materials Science and Engineering, 292*(1), 012070 (IOP Publishing).

Parra-Cabrera, C., Achille, C., Kuhn, S., & Ameloot, R. (2018). 3D printing in chemical engineering and catalytic technology: Structured catalysts, mixers and reactors. *Chemical Society Reviews, 47*(1), 209–230.

Schiffer, Z. J., & Manthiram, K. (2017). Electrification and decarbonization of the chemical industry. *Joule, 1*(1), 10–14.

CHAPTER 6

3-D printing in the construction industry

Contents

6.1 What?

At the end of the 20th century, with the development of programmable electronic technology, new production technologies also emerged. In particular, technologies for manufacturing new parts, products that do not require processing, and removing material layers have emerged. This new production technology provides a process that is in stark contrast to the traditional process of by applying materials layer by layer, including plastics, ceramics, and metal powders, to obtain parts or products. These parts are complete copies of 3-D models drawn on a computer. This technology is called 3-D printing or additive manufacturing in Western countries (Wu & Wang, 2016).

In the 21st century, the range of materials used for 3-D printing has increased significantly, originally making paper products and plastic products, but now parts can be made from metal, concrete, glass, and biological materials. One can create a full-featured product composed of multiple independent parts and components. This 3-D printing technology makes it possible to obtain parts and products without additional machinery and heat treatment, thereby significantly shortening the technical cycle time for product manufacturing. The principle of using 3-D printing to form shapes is called addition.

3D Printing
ISBN 978-0-323-99463-7
https://doi.org/10.1016/B978-0-323-99463-7.00006-2

First, a computer model of the object of the future is created either using a 3-D graphics editor CAD system or scanning the entire object in 3-D. Then, using a special software product, the printer divides the scanned object into several layers. A set of commands are generated to determine the order in which material layers are applied when printing. In addition, 3-D printers form objects layer by layer, gradually applying materials. The print head is positioned in a system with two coordinates X and Y, and applies materials layer by layer according to the analog electronic circuit (Attaran, 2017).

***From** (Source: ennomotive.com (https://www.ennomotive.com/3-D-printing-and-its-application-in-the-construction-industry).*

6.2 Where?

Recently, these 3-D printing technologies have been most widely applied in the manufacturing industries. They are particularly popular in the manufacture of simple polymer-based products, such as plastic tableware, toys, and interior items. Nevertheless, in the construction industry, the introduction of 3-D printing into existing building and structural technologies is in full swing, and so this innovation is considered in more detail here. In the early 2000s, several independent teams of scientists began to develop 3-D printing technology in the field of architecture. This innovation was encouraged by several factors, such as:

1. The global acceleration in various fields of human activities requiring the construction industry to drastically reduce the construction time of buildings and structures;

2. The continuous growth of the global population meaning the number of houses needs to be increased;
3. The existing construction technology incurs many material costs.

Therefore, there is a need to develop a new construction method to shorten the construction time, increase productivity, and reduce the cost of constructing buildings and structures. Although 3-D printing technology is relatively new in the field of architecture, it is now already clear that these innovations will soon become an indispensable part of human life. The most advanced countries in the development of 3-D printing technology in the construction sector are the United States, China, the United Kingdom, and the Netherlands. The results achieved in these countries are discussed in more detail in the following section.

6.3 When?

In 2015, at Loughborough University in the United Kingdom, a group of scientists obtained a unique cement-based mixture that has become an important raw material for printing plants of almost any type. This research team developed a patented 3-D concrete printing technology for the manufacture of full-scale construction and architectural components. The technology itself involves fusing this material layer by layer, which greatly simplifies the production, because the template is no longer needed. The resulting product can be easily adjusted and can also serve as a good basis for the finished work. Scientists at the University of Southern California (USC) went further than their British colleagues and suggested using 3-D printers directly on the construction site. 3-D printing has also entered the joint field of architecture and construction, from the realization of models and prototypes to the actual construction of life-size buildings. Professor Behrokh Khoshnevis at USC is developing a technology that will enable a house to be built within 24 h, including conduits for electrical, plumbing, and air-conditioning (Hazle, 2014).

3-D printing allows the creation of objects of different sizes, shapes, and materials, starting from a three-dimensional digital model, using special software design. The model is processed and then created by a special printer, working mainly by creating successive levels of layers, thereby overlapping the continuous cross-sections of the object to be copied. Chuck Hull can be considered the inventor of 3-D printing because he invented stereolithography in 1986. Today, these machines can print different materials, or even a combination of materials, to obtain the required physical and mechanical properties. Due to the general reduction in prices, this technology has become increasingly widespread, so that there are ever more affordable amateur 3-D printer models. 3-D printing also has applications in various fields, including prototyping and applications, such as jewelry, aerospace, industry, and medical equipment.

6.4 Why?

We live in an era where digitization, technology, and efficiency continue to drive the search for new solutions, which are often very innovative. This is the case of 3-D printing, which has great advantages, such as greatly reducing construction time and costs in the construction field. The construction time is shorter, and less construction personnel are required. Using a 3-D printer, one can create components for buildings, such as bricks, and create masonry directly on site. 3-D printing still has several key points that need to be addressed, including safety issues. Before using a building material, it is necessary to carry out all the performance and resistance tests that are usually performed on building materials, and if not impossible, they are certainly more complicated in this context.

In addition, there is currently a complete lack of shared and recognized standards and rules. Under current conditions, the most interesting application of 3-D printing in architecture is related to housing emergencies. In a short period of time, its direct on-site and low-cost construction are the characteristics that make it perfect to respond to emergencies caused by natural disasters.

From *(Source: 3-Dnatives.com (https://www.3-Dnatives.com/en/3-D-printing-construction-240120184/).*

For the same reason, 3-D printing technology is also being tested in developing countries. These countries need effective solutions to ensure people's health and adequate buildings, but at the same time reduce costs. The impacts of these buildings on the environment may be minimal, especially because today one can start printing from a mixture of locally available materials (such as cement powder, adhesives, and aggregates).

6.5 How?

3-D printers allow one to copy 3-D objects from digital models created using 3-D modeling software. This is a very fast and reliable form of additive manufacturing that can realize a great dream: materializing of any object in real time. This technology can print and assemble parts composed of various materials with different physical and mechanical properties in a single construction process. The object is created through a process that involves placing one layer on top of another and proceeding in cross-section.

In early 2011, at the dawn of this invention, 3-D printing enabled the creation of a single object as economical as the creation of thousands of objects, thereby destroying the economies of scale. Its impact on the world may be as far-reaching as the emergence of factories. Just as no one could predict the impact of the printing press in 1450, the transistor in 1950, or the steam engine in 1750, the same applies to this incredible 3-D printing technology. Predicting the long-term impact of 3-D printing, it is likely to impact every field that it touches in the next century.

There are different types of 3-D printers, and the most popular 3-Dwork through an additive manufacturing process, creating objects one layer at a time, starting with those below and gradually overlapping them until they cover the entire height. In most cases, the process can be performed by selective laser sintering, that is, by heating special materials, usually metal powders or thermoplastic substances, and then placing them in the correct position. Fused deposition modeling can achieve the same result, where heated nozzles are used to increase the temperature of the material before deposition. Filament, plastic or metal, is used and wound in a kind of strand that is gradually unrolled during the printing process. Another technique commonly used in the industrial field is object lamination production. Each layer is laser engraved and then positioned on the layer below.

From the beginning, 3-D printers have been used the industrial field. Using this technique, an object can be created quickly and cheaply, including prototypes. In this way, engineers and designers can see their work without having to start the entire production process. In recent years, the use of 3-D printing has expanded to various application areas: in the field of archeology, the technology is used to make 3-D models of only partially preserved discoveries or buildings, or integration of missing parts, or to create acquisitions through 3-D scanning a copy of an original object.

6.6 Who?

One of the most famous 3-D printer construction projects is a steel bridge over a canal in Amsterdam, the Netherlands, which is scheduled to be April 2018 completed. For this project, two printers are planned to be used, with each printer located at opposite ends of the bridge, working from the corresponding shore until they both reach the central area. Among the various applications of 3-D printing, architecture is also very important. In this field, 3-D printers are used today not only to create plastic models, but also to construct entire buildings. 3-D printing in the construction industry can save a great deal of energy and money, as it reduces the use of materials and labor. In 3-D printing, even a series of houses of different designs could be built in one go, including electricity, sewage, and air-conditioning pipes. Using ink formed from high-temperature concrete and fiberglass, multiple layers of reinforced concrete have been created to construct structures that workers can later assemble directly on the construction site.

The risks that the use of 3-D printers in building construction may pose to the landscape and construction sector remain to be assessed, but of course, this method has great potential in housing emergencies caused by natural disasters, or in extreme cases where conventional construction methods cannot be used. This type of 3-D architectural printing technology is very interesting for NASA because it may allow the first settlements on Mars and the Moon.

3-D printing is playing an increasingly important role in many production areas. The materials available for the use of 3-D technology have increased significantly and cover many applications. Technological innovations related to 3-D printing do not stop at the printing of plastics or metals. The cases of building structures (such as bridges and houses) made using 3-D technologies that work directly with concrete are not isolated. In this case, 3-D printing has also been recognized as a good method to build and reduce labor costs in a short time. In addition, with the use of additive technology, material waste is significantly reduced because only the amount required to execute the project is required.

Similarly, the systematic use of 3-D printing in buildings remains a huge challenge today. The first project aimed at using 3-D printing in the construction sector dates back to the 1950s, when automated machines using isocyanine foam and concrete were used. Later, other technologies were developed, such as D-Shape. ESA (European Space Agency) is particularly interested in this technology, as the agency aims to use it to establish a base on the Moon. The growing interest in 3D printing has made it possible to create new processes that bring to life the incredible pieces of modern architecture.

A 3-D printer, like a tower crane, is placed in a foldable state and installed on site after a complex earthwork. In addition, the production of all load-bearing elements of the building is carried out in the following way: the concrete pump delivers the ready-made mixture to the printer, and the printer applies it one by one to all the future load-bearing element buildings. This kind of work is very complicated, because it is carried out in the open, so it can be affected by unfavorable external factors. However, scientists at the University of Southern California have applied for a patent for a printer project called Contour Crafting at the US Patent Office. On this basis, these scientists plan to produce a large-size printer that can not only print the supporting structure of the building on site, but also print the required electrical and plumbing covered components.

In the field of 3-D printing of building structures, Eastern engineers are not as experienced as Western engineers. A printer named WinSun has been assembled in Shanghai WinSun Decoration Design Engineering Co., Ltd. This printer is large in size, measuring 150 m in length and 10 m in width. The productivity of this equipment is also amazing: it only takes a few hours to build a small building as high as 6 m. The raw material used is a glass fiber-reinforced cement mixture. The economic efficiency of this kind of production can be demonstrated by the fact that the cost of each house produced is about half as much as the cost of using traditional methods (Khan et al., 2020).

Also note that in the United States the series production of small houses using 3-D printers has been established. A private company engaged in this scientific research is led by the young engineer Andrei Rudenko. Andrei plans to develop and manufacture printers that can work on difficult construction sites: steep slopes or extreme weather conditions. There is already a small symbolic castle in Minnesota, proving that Ludenko's plan is viable (Tay et al., 2017).

Scientists from the Netherlands have adopted different methods to solve the problem of printing structures. They believe that it is impractical to print an entire building on a construction site. Instead, they recommend using a 3-D printer to produce special blocks. Such products are called Polyblocks. These are designed in such a way that the laying technique does not require a wet process. They are held together by special locks under the influence of gravity. A powder 3-D printer called ZCorp 510 is used to produce the agglomerates (Khan, 2018).

References

Attaran, M. (2017). The rise of 3-D printing: The advantages of additive manufacturing over traditional manufacturing. *Business Horizons, 60*(5), 677–688.

Hazle, M. (2014). *3D-Printing a building in 24 hours*. USC Viterbi, School of Engineering.

Khan, M. S. (2018). *3-D printing in transportation: Already in action* (Vol. 314). TR News.

Khan, M. S., Sanchez, F., & Zhou, H. (2020). 3-D printing of concrete: Beyond horizons. *Cement and Concrete Research, 133*, 106070.

Tay, Y. W. D., Panda, B., Paul, S. C., Noor Mohamed, N. A., Tan, M. J., & Leong, K. F. (2017). 3D printing trends in building and construction industry: A review. *Virtual and Physical Prototyping, 12*(3), 261–276.

Wu, P., Wang, J., & Wang, X. (2016). A critical review of the use of 3-D printing in the construction industry. *Automation in Construction, 68*, 21–31.

CHAPTER 7

3-D printing in dental care

Contents

7.1 3-D Introduction

Three-dimensional (3-D) printing is a comparatively modern production technology. 3-D printers utilize a computer program to lay down thin layers of product, such as powdered plastic or incinerated stainless steel, until the product has been formed. This innovation is now operating in many fields, from allowing astronauts to print instruments in space to enabling architects to build models. Its revolutionary dental care capabilities are numerous, such as personalized 3-D printed teeth, dental implants, and a number of other types of equipment for patients and dentists. The idea of 3-D printing has had huge impacts in many areas, and its potential uses are infinite, with especially exciting developments in the medical and dental field. 3-D printing remains a relatively young technology, but conventional dental labs could follow the direction of typewriters and film cameras if 3-D printing proceeds at its present speed. 3-D printing is becoming increasingly

3D Printing
ISBN 978-0-323-99463-7
https://doi.org/10.1016/B978-0-323-99463-7.00007-4

popular every day as technical advances begin to extend its use to a broader variety of applications, including several forms of health care (Attaram, 2017).

New technologies are enabling real achievements in the field of dental health. This is because, within this field, implantology is one of the most common and sought after treatments among patients. For this reason, technological advances have allowed a real revolution in this area. Currently, there are 3-D printers that allow the design and manufacture of dental prostheses. This is a great medical advance that has managed to speed up this procedure and also increase the precision in the results of the treatments.

3-D printing opens the doors to the future, and is turning science fiction into something real and present. With the introduction of innovative CAD-CAM technologies, the image of the dental sector has been changing drastically, and 3-D printers will further enable these changes. 3-D printing systems are capable of building 3-D models in a short time. Depending on the type of machine that is used, and the size and number of models that are being manufactured, the time required can vary significantly.

Rapid prototyping, also more popularly known as 3-D printing, is an additive manufacturing technology. 3-D printing is not new to the dental sector, in fact, it has already been incorporated into the production process in dental laboratories, specifically with the additive manufacturing system of laser sintering as it is popularly known, and the incorporation of other systems such as SLA, FDM, or PoliJet that have been introduced gradually.

The process begins with a design based on working models obtained using a laboratory scanner or intraoral scanner, and from there modeling software or computer-aided design (CAD) can be obtained from the design. The 3-D printing equipment reads the data from the CAD design and generally applies successive layers of liquid, powder, or thread material, manufacturing the physical model from a series of cross-sections. These layers, which correspond to the virtual cross-section of the CAD model, are automatically joined to create the final shape.

3-D printing uses a standard data file, popularly known as an STL file, to translate the data from CAD software to the 3-D prototyping equipment. The STL file contains the structure information using triangular geometric shapes or a mesh of triangles. In addition, this type of technology has achieved great popularity, in such a way that it allows the same design and manufacturing capacity that was previously exclusive to a few designers and large production companies.

7.1.1 Who had the idea of 3-D printing?

The general concept and methodology to be utilized in 3-D printing were first presented by Raymond F. Jones in his article "Tools of the Trade," released in the November 1950 issue of *Astounding Science Fiction* magazine. In that story, he alluded to it as a "molecular spray." Charles Hull, an American engineer developed the world's first 3-D printer in 1984. A regenerative medicine company called Organovo released data on the first fully bio-printed blood vessels and highlighted bio-printing technology in 2010 (Kessler et al., 2020).

The term CAD/CAM, applied to the dental world, refers to a technology that allows us to perform dental restorations using computer design support, and an automated machining or milling system that works at the user's command. It was introduced in dentistry in 1971, being at first more experimental and theoretical than clinical, and always focused on the field of fixed prosthesis.

The technique used in the first 3-D printer was stereolithography, considered a relatively expensive commercial technique, however recently low-cost desktop stereolithography printers have been introduced in the industry.

7.1.2 How do 3-D printers work?

In concept, 3-D printers are somewhat close to home inkjet printers that most people are acquainted with when printing papers and images from desktop computers. However, instead of pigment, 3-D printers use various plastic products and composite materials as raw materials. Although inkjet printers use a single layer of paper to create a printed image of the documents, 3-D printers add layer after layer, one on top of the other, plastic or composite in a very accurate series to eventually construct a 3-D replica of a computer-modeled material. This method is referred to as "additive manufacturing." If the current uses of additive manufacturing are already revolutionizing different sectors, its future possibilities in some fields, such as the health industry, could be remarkable. This wide technological variety already gives a clue that the applications of additive manufacturing are limitless. The initial utilization of rapid prototyping focused on preproduction models and in the manufacturing of finished products such as medical implants, automobiles, and fashion products prove that this type of manufacturing can open new gates and modernize the field of manufacturing and bring a positive impact to humanity. 3-D printers can produce almost any item that can be produced using a computer and fit into the printing compartment of the printer. Some 3-D printers are lightweight enough to work on a laptop and can create smaller items, while others are considerably larger and more suited for commercial or industrial purposes. 3-D printing has been increasingly embraced by many companies, as the pace and ease at which product models can be made have greatly increased (Sawhney & Jose, 2018).

Dentists, orthodontists, and oral surgeons use 3-D printing equipment in medical care as well as in study and research. Accurate replicas of almost every part of the human body, such as the jaw and teeth, can be made using revolutionary scanning techniques.

Three major steps are involved in the process of 3-D printing: procurement of data, processing, and actual fabrication utilizing additives. Data procurement is considered as the most crucial, as it requires images obtained from an MRI or CT scan. These images are then used to formulate a computer-aided design for the object, which needs fabrication. The coordinates from the 3-D model are used by 3-D printers to enable servo motors, which mobilize the printhead along the x—y plane, assembling a layer. The same movement is repeated along the z-axis for fabrication of the third dimension. STL file format is used for other data depicting texture, color, and thickness. Different types of materials are used for 3-D printing, such as stereolithography materials, acrylonitrile-butadiene-styrene (ABS), polyamide, polylactic acid, silver, titanium, steel, wax, photopolymers, glass-filled polyamide, and polycarbonate (Van Noort, 2012).

There are three basic stages in the 3-D printing process as follows. The initial step is to create a 3-D object file with a 3-D scanner using a CAD program. The second stage is the actual printing phase. The product to be used for printing is chosen as needed. Various materials used in dentistry include plastics, ceramics, adhesives, and metals. Most of these substances allow for a wide variety of finishing choices, thereby allowing a specific design to be accomplished. However, some substances are still under production. The third and ultimate finishing process necessitates special skills and materials.

7.2 How have dentists used 3-D printers?

Use of this technique in the healthcare sector didn't take too long, as scientists and engineers in the industry were able to start working with 3-D printing capability. Patients around the world have also begun to obtain

3-D-printed items to aid their well-being, such as prosthetics or bone replacements. Medical professionals benefit from the ability to easily plan and build personalized and advanced surgical instruments, allowing them perform their tasks faster and more effectively. Dentists, orthodontists, and oral surgeons use 3-D printing techniques in medical care as well as in study and analysis. Accurate replicas of almost every portion of the human body, such as the jaw and teeth, can be made using modern scanning techniques. By having such accurate replicas to work with, dental implants and corrective devices will be built to suit better, look better, and work better than ever before on each particular patient. Materials that can be used to build 3-D printing objects are progressing just as fast. Although most of these items are printed using plastics, a broader range of composite materials is now being used. Such composites are mixtures of materials used to create properties that are ideally tailored to the application in which a 3-D-printed product is used (Kessler et al., 2020).

7.3 Capabilities of 3-D printing in dentistry

Repairing or restoring a chipped tooth: The dentist scans the patient's mouth with a handheld optical wand. This produces a 3-D representation of the teeth and gums that is processed as a digital file. A computer-aided design (CAD) program allows the dentist to digitally design a dental patch and print the final result on a 3-D printer (Sawhney & Jose, 2018).

Restorative dentistry: Restorative dentistry is another area of dentistry. It has been enhanced tremendously due to 3-D printing innovation. The uses of 3-D printing are as wide as the strategies and procedures associated with restorative dentistry. From the production of printing trays to the final restoration, 3-D printing holds potential.

Building an orthodontic model: Pre-3-D printer technology meant making a patient bite down on a sticky, unpleasant, and messy clay so that it would harden into a mold that became the primary template for the construction of braces or Invisalign procedure. This is not the case for 3-D printing. The same equipment illustrated in the first case can be used by a dentist to scan teeth, layout an orthodontic implant, and print the final outcome in-house.

Manufacturing crowns, bridges, caps, false teeth: The same method as described above can be used for 3-D printing of all sorts of dental implants. The only difference is the same product used in the printing process.

Creating surgical tools: Not only can 3-D printers accommodate dental implants themselves, they can also print 3-D drill manuals used to complete those dental treatments.

7.3.1 Major practices implemented in 3-D printing

Light cure resins are used in digital light processing, photopolymer jetting, and stereolithography. The objects are built upside down through digital light processing by utilizing a light source and liquid resin on an elevating platform. In photopolymer jetting, layers are built incrementally as an ink type print head jets the light-sensitive polymer onto a platform. Meanwhile, successive layers are built by through stereolithography using a light-sensitive laser and light-cured photopolymer resin. The major disadvantages of these processes include, high cost, limited shelf life, and the fabricated products cannot be heat sterilized (Van Noort, 2012) (Fig. 7.1).

Another technique, referred to as the power binder is a process, which utilizes water through an inkjet printer head projected to a power bed for construction of an object. This technique is considered as cost effective, but has low-resolution output, greater operational difficulty, and hindered hear sterilization.

Technologies, such as electronic beam melting and selective laser sintering, sinter powder in a heated chamber below its melting point, followed by fabrication of the object by a scanning laser. Objects can be printed by selective laser sintering from metals and polymers. This technique is also considered to be relatively expensive due to the high maintenance and capital expenditure costs. It is also linked with health risks due to accidental explosion or dust inhalation.

Figure 7.1 Industrial SLS apparatus.

In fused deposition modeling, a nozzle releases small beads made from thermoplastic material for the construction of a model. 3-D printers using this technique have the highest utilization even at the domestic level and are often described as "home printers."

7.3.2 Applications

Medical modeling is considered as one of the earliest implementations of 3-D printing in surgery. It has been made more accessible by another technology which has become more conventional in recent years in the field of dentistry. In dental practice, Cone Beam Computed Tomography (CBCT) has become widely available and has brought about a transformation in diagnosis and treatment in endodontics and implant dentistry. Easy access to CT or CBCT in hospital settings means that volumetric image data can be easily provided to a 3-D printer before surgery and aids in making more detailed replicas of patient's jaws. This method allows complex anatomy to be reviewed carefully and facilitates the surgical approach to be planned before surgery. This has therefore led to the advent of new approaches and procedures in surgery. Moreover, the use of 3-D-printed technology may lead to less invasive, expedited, and more plannable surgery (Kiarudi et al., 2015).

7.3.3 Drilling and cutting guides

Drill guides are more commonly used in implant dentistry. These tools must be precise and robust as well as capable of disinfection and sterilization in a surgical environment. This technology is widely embraced by orthopedics also for total knee replacement. The use of drilling and cutting guides helps to formulate a virtual 3-D plan, made on a screen in any software, which is then transferred to the operative site. It is mandatory to

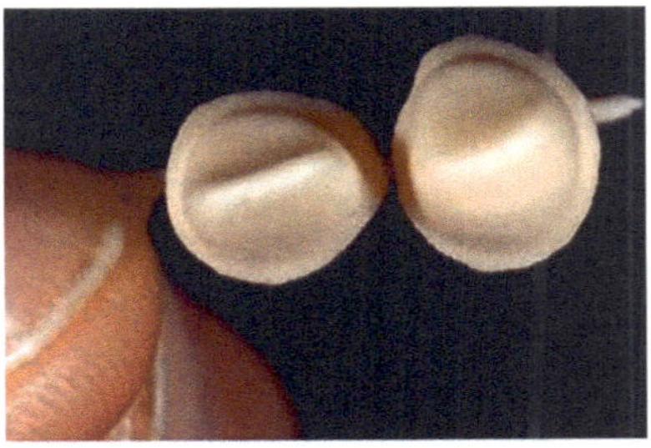

Figure 7.2 SLS-printed prepared teeth, printed from data from an intraoral scanner.

use high-resolution printing materials and precise 3-D printers for implant drill guides (Andonović & Vrtanoski, 2010) (Fig. 7.2).

7.3.4 Crown copings

A precise virtual model of the implant position, prepared tooth, and dental arch can be possibly developed by using laboratory scanners or intraoral optical scanners. In moveable and fixed prosthodontics, restorations may be designed and treatment may be planned using CAD software. These data and the CAD design may further be used to print or mill a crown, implant copings, bridge structures, and implant abutments (Andonović & Vrtanoski, 2010).

Fabrication of metal structures can be accomplished through 3-D printing, either directly in metal alloys or metals, or indirectly through the lost wax process by printing burned-out wax or resin. Moreover, direct printing in metals also poses specific safety and health requirements.

7.4 Benefits of 3-D printing in dentistry

Dentists save money: Adding to the dental lab is a substantial expense to any dental practice. The employment of trained workers in the manufacture of dental implants constitutes a significant ongoing expenditure. In all, based on the quantity of operations, the expense of operating a dental lab will approach thousands of dollars. Relate this to the one-time expense of around $20,000 for a top-of-the-line 3-D printer, which requires a starting stock of supplies.

Patients save money: the massive price of adding and maintaining a dental lab is transferred to the bills of each patient. Using conventional equipment, a mere crown will cost a patient approaching $2000 or perhaps more. As 3-D printers reduce operating costs by 80%, dentists can pass on the cost benefits to their customers.

Dental and orthodontic facilities are easier and more reliable: Manual modeling is time-intensive, while 3-D printing enables several devices to be produced at once. Precision is also enhanced when 3-D printers transform digital photos into real items by printing 16-micron-thick layers on top of each other. Enhanced processing capability and more reliable final outcomes help both dentists and patients.

7.5 Use of 3-D printing technology in endodontics and periodontics

7.5.1 Endodontics

Endodontics also profits from 3-D printing technology; in the manufacture of detailed manuals for use in both surgical and nonsurgical endodontic treatments. 3-D-printed surgical manuals assist with direct apicoectomy. Manuals in nonsurgical endodontic procedures are particularly helpful for the preparation of accessible cavities in calcified canals. The use of 3-D printing allows the development of dental prototypes of accurate anatomical root canal designs using CT images; thereby supplying dental students with the ability to work with realistic operations rather than the use of perfect typhodontic teeth.

7.5.2 Periodontics

3-D-printed manuals are most widely used in periodontology, for the aesthetic restoration of the gingival area. Detailed patient surgical guides are used for gingivectomy and smile design that make for accurate and personalized treatment. The application of 3-D printing techniques in regenerative periodontology is currently under review. Studies are being performed to test the use of 3-D-printed biphasic biomaterials for reconstruction and healing.

7.6 Advantages of 3-D printing technology in dentistry

1. Time efficient;
2. Precise specifics and procreation of the scan, giving good-quality work and accurate results;
3. Complex geometric patterns and interlocking portions that do not require assembly can be printed;
4. Significant decrease of production-related material losses;
5. It is possible to produce single objects in limited amounts at a low cost and to deliver them quickly.

7.7 Disadvantages of 3-D printing technology in dentistry

1. High cost of investment;

2. The final part is perhaps the greatest weakness in 3-D printing. Due to the manner in which each successive layer is mounted on top of the last in traditional 3-D printing processes, and intrinsic flaws can be clearly designed into the architecture.
3. Finishing of the end product is time intensive and needs specific skills;
4. Depending on the material, additional material could also be required to gain maximum output;
5. Stereolithography can only be achieved through light-curable liquid polymers;
6. The resin used is messy and can induce inflammation and discomfort through touch and ingestion;
7. The resin cannot be heat-sterilized.

3-D printing is the most revolutionary technological breakthrough that has arisen in the health industry in recent years, and it has tremendous potential. Dental patients should look forward to quicker care, faster healing time, and improved outcomes due to the fact that each 3-D-printed object has been specifically developed and uniquely built to fulfill the specific requirements of each particular patient. This extent of versatility and usability has never been so easy to reach in such a limited period of time, and things are only going to improve. The advent of digital dentistry has not only made treatments less time-consuming but also streamlined the approach to providing patients with a greater quality of life. While the initial cost is high and demands instruction in its use at the same it tends to minimize the time taken for direct medical treatment (Sawhney & Jose, 2018).

Digital technologies support patients by delivering services from all parts of the world by facilitating the sharing of information, and thereby enhancing healthcare overall. 3-D printing offers a better training opportunity for most dental practices, not just for clinicians, but also for dental students. It has immense promise for further testing and treatment. While it is not a substitution for traditional clinical approaches, the scope to strengthen and evolve is what this aspect of the horizon of dentistry offers.

In the field of dentistry, 3-D printing has diverse applications and holds an enormous amount of promise in making possible many innovative approaches and treatments for manufacturing dental restorations. Although no guidance in the use of 3-D printing technology has been implemented by national regulatory bodies, there is a dire need for regulators to set appropriate standards in this regard.

By introducing milling technology, a large number of options are now available for producing restorations, likewise, new materials for dental restoration to be used in 3-D printing are being developed regularly. Further conclusions can be drawn that slow evolution digital technologies and their use in dentistry has gained momentum to a level where more mainstream opportunities for the use of 3-D printing technology are presented in dental laboratory, orthodontic, and surgery areas. In addition to the affordability of 3-D printers, materials, cost of running, maintenance, and the requirements for skilled operators must be considered carefully. The correspondence of scanning, CAD, milling, visualization, and 3-D printing technologies, along with creativity, makes this an exciting venture for dentistry.

References

Andonović, V., & Vrtanoski, G. (2010). Growing rapid prototyping as a technology in dental medicine. *Mechanical Engineering Science Journal, 29*(1), 31–39.

Attaran, M. (2017). The rise of 3-D printing: The advantages of additive manufacturing over traditional manufacturing. *Business Horizons, 60*(5), 677–688.

Kessler, A., Hickel, R., & Reymus, M. (2020). 3D printing in dentistry—state of the art. *Operative Dentistry, 45*(1), 30–40.

Kiarudi, A. H., Eghbal, M. J., Safi, Y., Aghdasi, M. M., & Fazlyab, M. (2015). The applications of cone-beam computed tomography in endodontics: A review of literature. *Iranian Endodontic Journal, 10*(1), 16.

Sawhney, H., & Jose, A. A. (2018). 3D printing in dentistry-sculpting the way it is. *Journal of Scientific and Technical Research, 8*(1), 01–04.

Van Noort, R. (2012). The future of dental devices is digital. *Dental Materials, 28*(1), 3–12.

CHAPTER 8

3-D printing in the drone industry

Contents

8.1 What

The drone market continues to expand. Consumer Electronics Association (CEA) predicted that the global drone market will grow to 5 million units by 2023. In addition, U.S. drone sales are expected to grow 65% in 2030 to 2 million units. This growth will continue and may reach $3 billion by 2025, 10 times the current level (McNally, 2017). The commercial use of drones is wide-ranging. For example, Amazon has adopted drones as robots for delivery and has begun to use them for nutrient distribution and surveillance on farms. In fact, the U.S. Federal Aviation Administration (FAA) is working on legislation to allow drones to fly, with as many as 2 million drones weighing less than 25 kilograms expected to fly around by 2023.

The primary use of drones is for delivery. Amazon, as mentioned earlier, is attempting to match its delivery system with drones to make deliveries faster. In terms of delivery, Google is also embarking on drone development, and has already conducted a drone flight experiment in Australia and released the video. Google seems to be considering not only delivery but also security and surveillance. Already in the United States, drones have begun to play roles such as nutrient and pesticide spraying by utilizing

3D Printing
ISBN 978-0-323-99463-7
https://doi.org/10.1016/B978-0-323-99463-7.00008-6

drones for farm management. In addition to regular delivery, pharmaceutical companies are considering the development of a drone fleet that delivers medicines and medical products exclusively, and this is likely to become popular for applications that increase the added value of the company.

The global aeronautical context is characterized by very strong competition. The aeronautics industry, whether it involves airlines or the military sector, is in constant search for cost reductions, weight improvements, and maintenance reliability, while improving the passenger experience and reducing the emissions of CO_2 for commercial aviation.

From 3-D scanning to the use of additive manufacturing, 3-D technologies are already revolutionizing the aeronautics industry, and this is only the beginning.

The advantages provided by additive manufacturing are not unique to aeronautics, but they have a considerable impact in a sector marked by extreme competition and the search for equal or improved performance at controlled costs, as well as a marked dependence on the fossil fuel market Sought for benefits include the following:

- Improvement of the "buy to fly" **ratio**;
- Mass gain;
- Financial gains (reduced fuel costs, limitation of uses of expensive alloys, optimization of aircraft interior fittings, etc.), even if the latter are not yet guaranteed at this stage of maturity of the existing technologies;
- Possibility of **great design complexity**, whether this be for flight parts or prototypes;
- Decrease in spare parts stocks;
- Increased longevity and reliability of parts;
- Reduction (**or even elimination**) of the number of parts to be assembled;
- Fast prototypes;
- Much more responsive maintenance;
- Creation of **numbered parts** to distinguish each of them by a serial number.

Another very important contribution, whether in civil or military aviation, is that it will be possible to increase production capacity by **printing parts on-demand directly in flight or anywhere in the world**, including on ships and aircraft carriers. This is what Mike Murray of BAE Systems described during the flight of the first combat aircraft equipped with 3-D-printed metal parts in 2014: "*You are no longer stuck in a place to manufacture these objects.*"

Finally, the aeronautics industry has even more to gain by combining **3-D printing and topological optimization**: additive techniques do not increase the cost of producing parts with complex geometries, unlike traditional techniques.

Since the product is printed layer by layer from the bottom to the top, the dimensional accuracy of the product in height can be significantly influenced by the settings that specify the height of the first and subsequent layers of material exiting the extrusion head. The default layer height settings for standard printing modes do not take into account the overall dimensions of the model in height, which can lead to a decrease in dimensional accuracy. For the purpose of assessing the accuracy of printing in height and studying the influence of the height of the layers on the dimensional accuracy, products were printed in the form of a cube with a nominal edge size of 10 mm.

By selecting the optimal combinations of the height of the first layer during printing and the remaining layers, it was possible to print products with dimensional deviations in heights of less than 0.1 mm. It has been experimentally confirmed that in order to achieve maximum dimensional accuracy in height, the following condition must be met: the sum of the heights of the first and subsequent layers of material should be as close as possible to the value of the nominal size of the model height.

The deviations of the shape and size of products are greatly influenced by the technological modes of operation of the extrusion heads of a 3-D printer, which are set through the settings of specialized software that generates the control G-code (G-code is the actual file that is used by the 3D printer.). An important task is to determine the optimal values of the settings that ensure the receipt of the most accurate products in terms of size and the creation of applications that perform a preliminary analysis of the geometry of the 3D-model and allowing automation of the selection of optimal print settings.

8.2 Why

Among the best-known examples of the use of 3-D in aviation, is the Airbus A350 XWB and its 1000 parts have been printed in 3-D from 2014. The manufacturer is also continuing down this path, producing titanium engine mounts and airframes as standardization for the Stratasys solution for the A350 XWB supply chain.

General Electric, which is investing heavily in 3-D, has met with exceptional success with its LEAP engines, designed with Safran Aircraft

Engines, which integrate 19 fuel injectors printed in 3-D. Thanks to additive manufacturing, the engine is lighter (25% less mass), providing a lifespan five times greater than its predecessors, and only requires one part compared to the 18 parts that previously had to be assembled.

3-D is not just about the parts that make up the plane; they can also be parts used during maintenance or inspection operations, such as cockpit control guards.

8.3 Where

Although drones are expected to expand rapidly in this way, their product development seems to be becoming more high-tech. The Dutch drone maker Aerialtronics has reduced its development costs by as much as 50% by partnering with the 3-D printer maker Stratasys. Aerialtronics is a manufacturer of high-performance drones for a wide range of applications such as agriculture and livestock surveillance, security surveillance, and video recording in entertainment.

Its biggest feature is the small function of 35 people, but it can be customized to meet the needs of all walks of life. As mentioned above, drones are used in a wide range of industries and applications, ranging from delivery and monitoring in the field of entertainment. In addition, the environment used can vary even for the same event/procedure, and it can differ greatly depending on where the monitoring and delivery are taking place and the location of those monitoring the drone input.

There are various needs for these applications, and it is advised to customize them accordingly. However, Aerialtronics has been successful in reducing the development time by 50% using Stratasys' uPrint SE Plus 3-D printer. The uPrint SE Plus is a high-performance 3-D printer made of ABS filament that has optimum functions for manufacturing the parts required for drones.

***From** (Source: Shermon, A., & Moeen, M. (2019). Zooming in or zooming out: Entrants' product usage breadth in the nascent drone industry. Working paper, University of North Carolina, Chapel Hill.)*

For example, a drone generates a large amount of heat from the motor, so it can only be used with a plastic material that has a high heat resistance and constant strength and weight characteristics. From this point of view, ABS Plus, which is the material used by the uPrint SE Plus, is the best choice. In addition, a the flight test is indispensable for drone development. Once the design and prototype are complete, flight tests are performed 5–10 times, and if any deficiencies or improvements are found, the prototype itself must be remade.

In such a case, a high-performance 3-D printer that can directly produce plastic parts is more suitable than injection molding. In particular, Stratasys develops not only 3-D printing technology but also materials, so it provides a great deal of support in a wide range of product development fields.

8.4 When

The production of drones is increasing rapidly. It has been forecast that the drone market will grow from a value of $14 billion in 2018 to over $43 billion in 2024. With this fast-paced industry growth, rapid innovation is the key to competitive success, and 3-D printing is ideal for facilitating this (Cai et al., 2020).

From FPV racing to disaster relief, merchandise delivery, and reconnaissance, as the use cases of both professional and enthusiast drones continue to expand, 3-D printing is increasingly crucial in producing drones with new levels of performance.

3D printing reduces lead times during product development, reduces manufacturing costs, and enables designers and manufacturers to design new, state-of-the-art 3D-printed drones that offer new potential applications for the technology. From design to manufacture, learn how to build a drone with a 3-D printer and a DIY (Do-It-Yourself) 3-D printed drone guide.

8.5 How

With 3-D printers, designers do not have to spend hours analyzing the potential risks and shortcomings of new design features. High-resolution 3-D printers allow new prototype parts for testing to be quickly created with minimal waste.

Drone manufacturers can manufacture replaceable modules with accurate geometric accuracy without the need for expensive and time-consuming molded parts.

Introducing a 3-D printer into the process reduces lead times by taking only a few hours to manufacture new parts. For example, the MRB-1 drone moved from the initial concept design to a working prototype in 4 weeks. 3-D printing not only offers the benefits of design freedom and time savings, but also leads to significant cost savings. CAD files that can be shared and easily modified during the design phase also make it easy to duplicate the design. This makes risk-taking and innovation when designing drones much more affordable. While developing its own drone, the US Marine Corps discovered that the drone manufacturing process would cost about $600, excluding the cost of purchasing a printer; compare this to a ready-made raven drone that suited their needs, which can cost up to $30,000 (Chang, 2020).

3-D printing makes it easy to manufacture replacement parts for a drone. For the Marine Corps, the cost of replacing a damaged wing on a high-performance drone, which previously cost $8,000, is now only 0.1% of that (Kozlova et al., 2021).

8.6 Who

Many drones use the popular quadcopter design, but the versatility of 3D stamp printing gives developers the freedom to come up with fairly ambitious designs and let their imaginations run wild. Winning the UAV Zelator 2016 Airbus Cargo Drone Challenge entrants are optimized to deliver small cargo, especially medical supplies, to remote locations. The Zelator has a fixed wing design. Creators use 3D printed materials to create a body that is strong enough to withstand inclement weather, yet light enough (under 25kg) to work in harsh environmental conditions may encounter. Ogasawara Yuki and Kumeda Ryo's X VEIN are 3-D quadcopter drones. Designed as a "lifesaving drone," the X VEIN is optimized for performance during disaster relief efforts. It boasts extended flight time, a reinforced frame, and a propeller guard to prevent damage from collisions. All of these features are built into the lattice-structured quadcopter body created by 3-D printing.

The high degree of customization in drone design, the complexity of the lattice structure, and the free adjustment of density within the generated

CAD software are all uniquely possible with the capabilities of the laminated molding technology. Many manufacturers have experimented with larger drones, but there is also design creativity at the other end of the scale. For example, UPenn's Piccolissimo is one of the smallest unmanned aerial vehicles ever manufactured, measuring 24.26 mm in diameter and weighing only 2.5 g. With a single propeller and chassis, the Piccolissimo is suitable for harsh conditions. Miniature size (24.26 mm in diameter) and ultralightweight (about 2.5 g) can only be achieved by using 3-D printed parts.

Due to its weight and size, this small drone is ideal for operations in volatile nuclear reactors and for reconnaissance in confined spaces and remote areas. Such tasks are impossible for humans and larger drones to accomplish. Piccolissimo also foresaw future successful use cases in the areas of 3-D printing and swarm robotics. Manufacturers can use 3-D printing methods to create tools like Piccolissimo for swarm robotics efforts such as rescue missions. Let's hope that technology has only benign uses, unlike in the notable episode of Black Mirror.

With 3-D printing, you can manufacture most of the components needed to build drones and quadcopters, in addition to electronics. 3-D-printed drone parts include chassis, additional protective casings for frames, propellers, landing gear, motor mounts, accessory mounts (for scanners, cameras, sensors, or other useful equipment), remote controls, and drone power supplies (Rayna & Striukova, 2021). To build your own DIY drone, you can choose from a variety of off-the-shelf open-source designs, including the OpenRC 3-D printable quadcopter design. Communities like 3-D model databases and Pinshape also offer a wide range of drone designs and accessories that allow full 3-D printing.

Finally, unique 3-D modeling of new parts and accessories offers endless possibilities for adapting your drone design to a particular application. RC groups, DIY drones, and of course, online communities like Reddit can also provide ample support and inspiration. To 3-D print a drone, either common plastic 3-D printing techniques, fused deposition modeling (FDM), stereolithography (SLA), or selective laser sintering (SLS) can be used, depending on the requirements. Of the three technologies, SLA 3-D printing is perfect for precision and use of a variety of materials. Due to the nature of drones, it is advisable to choose a material that can be processed without negatively affecting the compression, elongation, bending, and impact properties, such as the tough and durable family of SLA resins.

From *(Source: Fey, M. (2017). 3-D printing and international security. Peace Research Institute Frankfurt, PRIF Report (Vol. 144).)*

8.7 The specificities of the aeronautics sector

The aeronautics industry was one of the *early adopters* of 3-D printing from the end of the 1980s. Additive manufacturing was then used for prototyping and not for production.

But unlike other sectors with more flexible standards, aeronautics is characterized by a high degree of safety requirements which may have slowed down implementation until then.

The drastic aircraft standards are, of course, intended to ensure that no defective part causes an in-flight accident, but they also make it possible to limit the time for managing technical damage on the ground, the immobilization of an aircraft being particularly expensive.

The evolution of 3-D printing now makes it possible to overcome this delay and obtain the necessary certifications, even if these take several years in aviation. Today, aircraft manufacturers rather certify 3-D manufacturing processes for existing parts rather than original parts.

Finally, the aeronautics industry must also take up the challenge of R&D in terms of the materials to be used. The powders remain particularly expensive and still limited in composition.

8.8 Conclusion

The need and spread for drones will continue to grow in the future. Many drone makers have already appeared, and they are developing products according to their applications. In order to expand such drones, it will be most important to adapt them to all environments and various uses. Conduct usage experiments and flight tests for each environment, and recreate prototypes to adapt to them.

Such work is required. In addition, 3-D printers will play a major role in accelerating the development of such prototypes. The work of Stratasys and Dutch drone maker Aerialtronics is a perfect example. On-demand 3-D printing technology with a thorough knowledge of materials and functionality will play an important role in responding to more detailed user needs.

References

Cai, Z., Zhou, Y., Weng, Z., Deng, L., Luo, Y., Yu, M., & Qi, Y. (2020). 3-D printing Conformal K-Band Lens Antenna for a Smart Parking space Detection system. *IEEE Transactions on Instrumentation and Measurement, 69*(10), 8514–8525.

Chang, Y. N. (2020). *Lack of adoption of 3D printing in Malaysia construction industry*. (Doctoral dissertation, Tunku Abdul Rahman University College).

Fey, M. (2017). *3D printing and international security, 144.* Peace Research Institute Frankfurt, PRIF Report.

Kozlova, E. V., Yakinchuk, V. V., Starikov, K. A., Bolshakova, A. V., & Bocharov, A. A. (2021). January. Life Cycle management of the 3D-printer technology to design an Underwater drone Hull to study the Arctic. In, *Vol. 625. IOP conference series: Earth and environmental science* (p. 012016). IOP Publishing. No. 1.

McNally, D. (2017). *Army engineers demonstrate new system for on-demand 3-D printed drones.* US Army. https://www.army.mil.

Rayna, T., & Striukova, L. (2021). Assessing the effect of 3D printing technologies on entrepreneurship: An exploratory study. *Technological Forecasting and Social Change, 164,* 120483.

Shermon, A., & Moeen, M. (2019). *Zooming in or zooming out: Entrants' product usage breadth in the nascent drone industry. Working paper.* University of North Carolina, Chapel Hill.

CHAPTER 9

3-D printing in education

Contents

9.1 What

3-D printers in the education field are an excellent opportunity to cultivate spatial thinking and creative skills. Practical modeling has fundamentally changed children's views on various subjects, making the teaching process of programming, design, physics, mathematics, natural sciences, and other sciences easier to understand. In addition, creating something with your own hands helps to cross the threshold of passive consumption of ordinary goods that our society is accustomed to, and turn ideas into reality. Nowadays, buying 3-D printers for schools is no longer difficult. There are a wide variety of complex levels of equipment on the market, and the prices are affordable, for children and also for teachers. 3-D printers are no longer an expensive tool only for large companies and leading research centers. Today, even children can become users of 3-D printing devices, with prices being only a few hundred dollars. The use of 3-D printing technology is revolutionizing teaching methods and elevating the educational process to a whole new level of student creativity development (AbouHashem et al., 2015).

3D Printing
ISBN 978-0-323-99463-7
https://doi.org/10.1016/B978-0-323-99463-7.00009-8

Source: 3Dnatives.com (2021).

9.2 Where

3-D technology can be applied to all levels of education from elementary school to university, not to mention that almost any subject can be taught more easily using 3-D printing. For example, a physics teacher studying the principles of jet engines may ask students to create a working 3-D model of a rocket. A geography teacher may ask them to print a 3-D model of any part of the world to help the children better understand the area under study. This makes the learning process more interesting and the knowledge gained through this experience is more sustainable. Regardless of the level of education, there are several key points that should be considered when purchasing a 3-D printer for an educational institution:

- Safety. It must be remembered that very inexperienced users may experience the irresistible desire to touch everything; However it should be remembered that the temperature of the print head of the 3-D printer with FDM/FFF technology is 100–500°C, and the photosensitive resin is very difficult to clean. Therefore, a 3-D printer in a separate room would be a wise choice. Note that printers do not always only emit harmless vapors from the print media when printing, so educational institutions need them to be equipped with trap filters.

- Equipment cost. No matter what preventive measures the teacher takes, the possibility of equipment failure during the training process is very high, which can increase the maintenance costs. If it is a cheap device, it can be completely replaced or replaced by a larger unit.
- Simple design. The simpler, stronger, and more reliable the invention is, the less likely it is that some erroneous operations will cause expensive equipment to fail.
- Availability of consumables. Although you may want to use the very expensive branded materials to get the perfect product, you should remember the main purpose of buying a 3-D printer, which is for training. Therefore, there are many defective products, material consumption is always high, and the printer must be able to print the cheapest and most common of them (Buehler et al., 2016).
- Service: convenience of repairing and replacing faulty parts. Training is an ongoing process, and manufacturers' representative offices and service centers located nearby will save a lot of time and money in the repair process.
- Use free software. The initial training is not designed to turn users into professionals. Teachers should teach students the basic knowledge about 3-D technology and the desire to further develop these technologies. The use of complex professional software packages, in addition to their high cost, usually makes inexperienced users experience self-doubt.
- Brand training courses. Many 3-D equipment manufacturers interested in advertising their products have created special training courses for students and teachers to use their equipment. The existence of such a system will enable the avoidance of many errors and increase the effectiveness of training.
- Brand incentive plan. Some companies have funding programs to motivate the most talented teachers and students.

Universities and research institutes improve the knowledge acquired by students in schools and provide professional knowledge for mastering professional knowledge. In addition, they are also research and development centers for new technologies. The use of 3-D printing speeds up these processes. Using 3-D printing, medical students can print human organs to improve their knowledge of anatomy; technical students can quickly prototype new products; and architecture and design students can publish their models (AbouHashem et al., 2015).

9.3 Who

Schools and teachers who want to integrate 3-D printing into the classroom must obtain financial or practical support. Although many children do not know how to use 3-D printers, the same can apply to teachers. They need to receive technical training and acquire the necessary skills, and then pass them on to their students. This is why there are networks, grants, and other programs around the world to support the implementation of 3-D printing in schools.

In 2013, the British Ministry of Education allocated more than £500,000 to enable 60 schools to be equipped with 3-D printers and to train teachers to use them. Financial assistance makes it easier for these institutions to integrate 3-D technology and provide young people with concrete and useful applications. Also in the UK, other initiatives are in place to promote 3-D printing in schools, such as an education project platform that connects companies specializing in 3-D technology with schools and encourages teacher training.

In Germany, Com3du is being developed in cooperation with Ultimaker to provide machines, accessories, educational materials, advice, and experience in classroom 3-D printing. This platform now consists of 50 member schools and academic partners.

9.4 When

The first step before starting printing is becoming familiar with the slicer program. The slicer software is usually provided with the 3-D printer. In the Zenith 3-D printer, the slicer is built into the control program of the Repetier-Host printer. With the help of the slicer program, a 3-D model can be prepared for printing with specified parameters. 3-D printers need clear instructions: how to move the carriage, layer height, printing material and temperature, how to fill internal voids, and so on. Therefore, novice 3-D printers (regardless of age) need to learn how to use this software product before starting to print directly. The second stage of 3-D printing training is training for interacting with 3-D printers: They need to understand the characteristics of various types of plastics (filaments), adhesion (or more simply, adhesion), the supply of plastics, etc. Taking the printing of 3-D models with different settings as an example, students can reach an elementary level of understanding the printing process and develop an understanding of the 3-D printer functions they can use.

9.5 Why

The use of 3-D printers is becoming increasingly widespread, and these machines have become common in companies, homes, and educational centers. There is no doubt that the use of 3-D printers in education offers great potential for students and teachers. 3-D printers are not yet common in education centers, although there are some experimental initiatives that have succeeded in introducing them. In countries such as the United States and the United Kingdom, they are included in the classroom, realizing their potential as an educational resource, their versatility, and the possibility of promoting active participation of students.

As computers and interactive whiteboards entered the classroom 10 years ago, then these types of printers are likely to be the next. They will become supporting tools for science, art, and/or history disciplines. A 3-D printer is a machine that can replicate designs in three dimensions. Its main value is that it can take the configuration stored in a computer file and turn it into the reality of a block or volume model with length, width, and height. The operation of 3-D printing is also called additive manufacturing. It can be printed without any molds or tools. Bit by bit, the machine builds the part by depositing layers of material, which solidify and form the desired structure. The construction process is progressive, from the bottom to the top. 3-D printers usually use thermoplastic materials that are melted by filaments and which harden when cooled. The main materials used are Acrylonitrile Butadiene Styrene (ABS) consumables and Poly Lactic Acid (PLA) consumables, which account for 95% of the materials used in this type of printing.

There are many advantages to using 3-D printers in education. First, the application of 3-D printing has increased student participation. They generate greater participation in the classroom because they transform learning into a different and interesting experience, and most importantly, make the subject matter more tangible. Schools usually use them at group activities to experiment with machines or hold practical courses where students can discuss and contribute their ideas.

Another reason for their implementation is that they promote interdisciplinary learning and teamwork. To make a topographic map, you need to understand not only geography, but also science or mathematics. Students will have to research and learn various topics to make their 3-D prints, which encourages collaboration between disciplines. Working methods are usually grouped, which means knowing how to work with others and respecting the opinions of others.

3-D printers are a resource to connect with current students (tech-savvies) and attract their interest. A 3-D printer is a resource that connects with current students (digital natives) and attracts their interest. This is a practical course with real effects and fast results, which makes students take it seriously, listen carefully to the teacher's explanation, and make their own works. In addition, they are a good resource to make certain topics easier and more dynamic, such as explaining how the pyramids of Giza were made in Egypt.

3-D printing can improve teaching and help clarify concepts. The educational applications of these printers promote the work of teachers, thereby improving the quality of education. They are most commonly used in the fields of mathematics, science, technology, and engineering, where they use abstract concepts and vague terminology. If real applications are realized, it will be easier to absorb them. However, their use should not be limited to science subjects. One can also use them in geography and history—making topographic maps, landscapes, historical buildings—or use them in music to create an instrument.

They improve creativity and problem-solving skills. Thanks to 3-D printers, ideas and projects can be turned into real objects. This helps to change the student's mentality, because it forces him to think, identify, and resolve possible setbacks. In many cases, students are forced to use creativity and seek solutions to achieve established goals. For this reason, they help to find mistakes and learn from them, because the student usually does not realize the failure of a project until he thinks it is "reified." According to a real example, copying the Burgos Cathedral model in a high school group, knowledge of architecture, history, and engineering were enhanced. Each student is expected do his best and work with his classmates, to realize their first big project.

There is no doubt that having a 3-D printer in an education center has a number of advantages: it opens up students' minds, they learn teamwork, and at the same time they acquire new working models that translate theory into reality. This is an innovative method, and the key lies in the role of the teacher. The teacher must coordinate and allocate the functions of each student and receive appropriate training to understand how to introduce new technologies in the classroom (So et al., 2017).

3-D printing used to be available only in industry, but in recent years it has experienced a strong democratization. Today, thanks to desktop 3-D printers, this technology has become easy to use and essential for education. 3-D printers are integrated into education by discovering innovative technologies, developing student creativity, and implementing educational and technological projects. Some countries are lagging behind in integrating new technologies into schools; the fact remains that students are affected by this digital and industrial transformation. 3-D technology should be included as an example of achieving this change. Many school subjects, regardless of the level—from middle school to university—can use 3-D printing to simplify concepts and help students understand certain theories. In chemistry, teachers can 3-D print representations of molecules; in SupraVentricular Tachycardia (SVT), various parts of the body can be easily modeled and then printed; and in history, we can create well-known monuments, and so on.

In addition, by integrating it directly into the classroom, students will become more familiar with the technology. This is important if we consider 3-D printing as the fourth industrial revolution. Digitization and technology are changing everything. Especially in schools, increasing digitalization is a major issue, and the field of 3-D printing is particularly important. Students must prepare for the future. But these changes are so profound that putting an iPad in the classroom is not sufficient; it is necessary to accompany them. The same is true when using a 3-D printer or 3-D pen.

9.6 How

The application of 3-D printing in medical education has endless possibilities. From printing molds to making in-ear headphones, to making prosthetics to replace human limbs, it has quickly found itself at the core of many hopes for improving the quality of human life. Other health fields are no exception. In plastic surgery and dentistry and even surgical operations, hopes are high: instruments, implants, models, prototypes, etc. can be manufactured using 3-D printing.

Initially, the integration of 3-D printers belonged to the STEM (Science, Technology, Engineering, and Mathematics) education model. These applications are mainly related to science and mathematics: using geometric formulas to better visualize complex shapes, manufacturing robotic parts, or printing models related to understanding how the human body works (Buehler et al., 2014).

The STEAM (science, technology, engineering, art, and mathematics) model rapidly enhanced the STEM model. Artistic applications are rapidly appearing now: they allow the creation of models of various different works and, of course, have become a new bastion of student creativity.

Source: makeblock.com.

Finally, the humanities are not excluded. In geography, it has become easy to print reliefs to simulate hydrological basins or conceptualize the development of urban areas. Historically, from the mud houses of the Vikings to the great European cathedrals, it was very simple to replicate the various tools used in ancestral techniques and build monuments based on different architectural features.

When the benefits of 3-D printing for all students are envisaged, 3-D printing in schools becomes increasingly meaningful. How to integrate it effectively into the classroom remains to be seen. Since 2013, the Alsace School in Paris has been integrating 3-D printing into the classroom. It regularly organizes seminars for students to familiarize them with 3-D technology. Children of this age are very creative. Unlike the prior, they imitated our "little hen" example in minutes. As long as the project is led by teachers and students are regularly guided and encouraged to use 3-D printers, 3-D printing in schools will make sense and increase employability (Buehler et al., 2014).

9.7 Conclusion

Education with 3-D printing can be started from an early age. Some schools integrate 3-D printing into the classroom and equip them with machines suitable for children. The number of students learning 3-D printing technology in school is increasing, because it can teach more about adapting to the industrial sector, architecture, and even design. However, when we carefully observe the practices in schools, we see that there some obstacles remain to adopting 3-D printing more widely. Several important factors must then be considered, first of all being the possibilities and advantages that 3-D printing provides for education. This technology must be implemented effectively, and young people should be encouraged to use it (Ford & Minshall, 2019).

Teachers must introduce students to specific applications of 3-D printing and show how to use them daily. This is also one of the most important obstacles to the adoption of 3-D printing in schools. It is not sufficient to install a 3-D printer in every school. There must also be one or two people who are familiar with this technology in each school (through training or enthusiasm), otherwise (as already happen in some schools) the printer will accumulate dust.

The advantages of using 3-D printing, 3-D printers, and 3-D modeling in education are obvious—they not only help to absorb knowledge in traditional disciplines, but also provide specific knowledge and skills that are much needed in our time, and help to continue education and application in practice. Aware of this, increasing numbers of educational institutions are turning their attention to digital and additional technologies, which are essential for students to prepare for adulthood. Seeing this trend, companies that produce 3-D devices are providing more and more products and programs dedicated to education.

References

AbouHashem, Y., Dayal, M., Savanah, S., & Štrkalj, G. (2015). The application of 3D printing in anatomy education. *Medical Education Online, 20*(1), 29847.

Buehler, E., Comrie, N., Hofmann, M., McDonald, S., & Hurst, A. (2016). Investigating the implications of 3D printing in special education. *ACM Transactions on Accessible Computing (TACCESS), 8*(3), 1–28.

Buehler, E., Kane, S. K., & Hurst, A. (October 2014). ABC and 3D: Opportunities and obstacles to 3D printing in special education environments. In *Proceedings of the 16th international ACM SIGACCESS conference on Computers & accessibility* (pp. 107–114).

Ford, S., & Minshall, T. (2019). Invited review article: Where and how 3D printing is used in teaching and education. *Additive Manufacturing, 25*, 131–150.

So, H. J., Lee, J. H., & Kye, B. (2017). An exploratory study about the activity framework for 3D printing in education and implementation. *Journal of The Korean Association of Information Education, 21*(4), 451–462.

Further reading

3 D printers for kids & STEAM education - makeblock.(2019). Retrieved 26 March 2021, from https://www.makeblock.com/official-blog/260877.html.

Is the implementation of 3D printing in education a necessity? - 3Dnatives.(2018). Retrieved 26 March 2021, from https://www.3dnatives.com/en/3d-printing-in-education-290820184/.

CHAPTER 10

3-D printing in the fashion industry

Contents

10.1 Introduction

3-D printing is a type of prototyping technique that is also known as additive manufacturing. As science and technology have developed, the technology of the fashion industry has entered the era of 3-D printing from the previous textile era. 3-D printing fashion includes all wearable items from clothes and shoes to accessories. Although the clothes produced by 3-D printing do not appear in clothing stores, it has brought great innovation to the fashion industry (Shahrubudin et al., 2019).

3D Printing
ISBN 978-0-323-99463-7
https://doi.org/10.1016/B978-0-323-99463-7.00010-4

10.2 What

The origin of 3-D printing technology began as early as the end of the 19th century. At that time, this cutting-edge technological idea was already understood and recognized in the United States. In the early years, this technology was called "rapid prototyping" and was mainly used in specialized industrial design and production. With the continuous development of this technology, 3-D printing has went deeper into various fields, so that people's awareness of it has gradually deepened, and more and more people began to recognize it. It was not until the mid-1990s that "rapid prototyping" was described as 3-D printing (Norani, 2017).

10.3 When

Throughout history, technological innovation has been an eternal topic, especially in the industrial age. Therefore, to encourage the development of new technologies, the fashion world has begun to vigorously promote product innovation, talent introduction, and the application of new technologies in various fields. 3-D printing technology has been maintaining steady development.

The use of the layered molding method promoted the beginning of 3-D printing technology, and the three-dimensional graphics produced using this method mark the beginning of the history of 3-D printing technology. The next step is to tentatively propose the use of light-curable materials to pile up the laminates until the final three-dimensional model is formed. Next is the method of laser selective irradiation of photopolymer, as proposed in 1977, to directly create a three-dimensional model (Yan et al., 2018).

10.4 Who

In 1979, professors from the University of Tokyo in Japan began to use thin-film technology to make tools. In 1982, they tried to apply optical technology to the field of rapid prototyping. In 1988, 3-D Systems highlighted the world's first SLA-250-based printer, known as the "three-dimensional plane printing machine." In 1989, a variety of materials were formed to diversify 3-D printing production. In 1992, the Stratasys and 3D systems companies introduced the first 3-D printer based on fused deposition modeling (FDM) technology. In 1993, MIT used adhesives to bond metal and other materials into shape. In 1996, rapid prototyping technology was accurately described as "3-D printing." In 2005, 3-D printers entered the color era. The year 2012 is memorable in the history of 3-D printing around the world, with *Popular Science* calling this innovation "3-D printing technology" for the first time.

10.5 Where

On April 19, 2013, according to foreign media reports, San Francisco-based startup Continuum was using 3-D printing technology to produce "fashion in the digital age" products. Continuum allows users to design ready-made garments online and print out these personalized garments using a 3-D printer. Until recently, 3-D printing was limited to large companies, but the cost of this technology has been greatly reduced in recent years, allowing new and small companies such as Continuum to use 3-D printing technology to produce their products instead of relying on factories (Vanderploeg et al., 2017).

3-D printing technology is not new, first appearing in the 1980s, and has had a direct impact on the fashion industry. Previously, after designers designed a piece of clothing, they had to be sure that this clothing piece would sell hundreds of pieces to earn back the manufacturing cost, but 3-D printing has completely subverted this idea. Before a consumer orders a certain piece of clothing, the designer's production cost is almost zero. Therefore, designers can now afford to create small-scale experiments and sell limited-edition garments. Big fashion brands such as Nike are already using 3-D printing technology and customized services. For many years, this shoe company has allowed buyers to customize the color and design of their footwear. March 3, 2013, Nike launched the world's first printed football boots. The 3D printed technology can customize football boots and offer more customized services (Kim & Kang, 2015).

Certainly, 3-D printing clothing is still several years away from entering mainstream production. One reason for this is that the cost remains high. Bikini tops on the Continuum website cost $250—300, and shoes are $900 a pair. Also, the finished products may have some defects. "Wired" pointed out that the plastic straps of the 3-D printed Continuum bikini turn yellow in the sun. However, this technology will continue to develop, and the cost will continue to decrease. Internet users can also now design for themselves on their own computers.

Continuum has a clothing design database from which consumers can download clothing design materials for free and print them out at home if they have a 3-D printer. Since 2006–11, the sales of 3-D printers have increased by 35,000%, so that this scenario will soon become the norm.

10.6 How

Today's 3-D printing technology is easy to use in the manufacture of shoe soles and jewelry, but 3-D for clothing is a more challenging process as it requires flexibility, as 3-D printing is mainly used to produce rigid, nonflexible parts. As clothing is the most challenging part, 3-D printing of flexible clothing has always been a challenge faced by designers due to its softness and "mobility" Although 3-D-printed clothing is still mainly used in fashion shows, there also have been some promising projects (Mpofu et al., 2014). The most unique products in the fashion industry are as follows.

10.7 Apparel: 3-D printing clothes by Danit Peleg

As early as 2015, Danit Pele launched her first fashion series Liberty Leading the People. The unique feature of Farrer's 3-D printed clothes is that they are all printed on a 3-D printer at home (Saphio et al., 2016).

10.8 Jacket without seams

The Ministry of Supply has launched a unique 3-D jacket made on a 3-D robotic knitting machine.

10.9 Jewelry example: nervous system bracelet

The 3-D silver bracelets are made on nervous system structures.

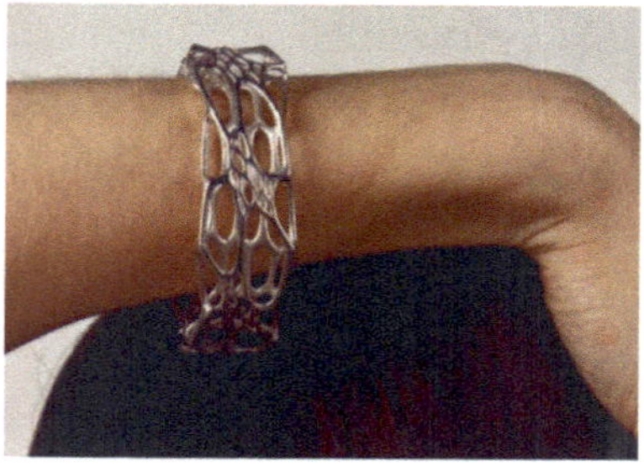

10.10 Footwear

By using Stratasys, designer Ganit Goldstein produced a 3-D-printed woven shoe, which can be printed in a variety of colors and materials. In addition to 3-D printing, Goldstein also uses Thai dyeing technology, and the illustrated pair of shoes was displayed at the 2018 San Francisco Fashion Art Competition on December 17.

10.11 Watches

The watch industry uses 3-D printing for the creation of prototypes, and also for general manufacture, as can be seen in one of the latest watch models from the Panerai brand "Lo Scienziato Luminor 1950 Tourbillon GMT Titanium," printed with the help of 3-D technology, by using the direct metal laser sintering (DMLS) technique.

10.12 Ornaments

3-D technology can be used in creating ornaments from traditional textiles, such as the 3-D-printed clothing modified by a young designer named "Anastasia Ruiz" who collaborated with Sculpteo. From buttons and rings to leather goods, 3-D equipment primarily allows the use of 3-D printing material made of thermoplastic polyurethane (TPU) and other polymers such as polyamide (Sedhom, 2015).

10.13 Recent 3-D fashions

Given the design freedom that additive manufacturing allows, it makes sense that designers would be interested in working with this technology to create complex shapes and original parts. From shoes to dresses, 3-D printed fashion is very different from conventionally manufactured clothing. Each design tends to be unique, personalized, and sustainable. This is another key as to why 3-D printing is of interest for this sector. Reducing waste in fashion is a challenge that some creators wish to address for their collections. In the following list, sorted in alphabetical order, we review some of the most innovative 3-D-printed fashion pieces (Valtas & Sun, 2016).

10.14 Annie Foo, original 3-D printed shoes

Annie Foo is a high-end shoe designer who uses 3-D printing technologies in her manufacturing process. She explains that she has always been fascinated by combining existing shoe-making methods with modern technologies and materials to create something new. And this is exactly what these 3-D-printed shoes are: originals. Using 3-D modeling software, she designs complex shapes to give her shoes an aesthetic look. Her final pieces, produced for the catwalk, were made on HP's Jet Fusion 4200 machine, using PA12 (nylon).

10.15 Anouk Wipprecht and her Proximity Dress

Anouk Wipprecht is a Dutch stylist known for combining new technologies and fashion to create unique garments, with unexpected characteristics. This is the case with her new dress called the Proximity Dress. Anouk envisioned it as part of the health crisis caused by COVID-19. The dress includes different sensors that can detect breaches of the mandatory social distance. If a person gets too close to the wearer of the dress, the dress stretches to create a kind of barrier. This is a clever idea developed in part thanks to 3-D printing: Anouk used the SLS process and Stratasys PolyJet technology to manufacture different components of the proximity garment.

10.16 Ganit Goldstein, custom fashion

For a few years, the young stylist Ganit Goldstein has been using 3-D printing to create bespoke clothes that perfectly suit everyone. Her first collection, called "Between the Layers," includes seven garments and six pairs of shoe 3-D-printed with Stratasys machines, in particular a PolyJet that allows the combination of multiple colors. Ganit Goldstein explains that she always begins her work with a 3-D body scan to adjust her creations to a specific silhouette. She hopes to offer custom, comfortable, and appropriate clothing for everyone.

10.17 Iris Van Herpen, between 3-D printing and nature

Iris van Herpen is undoubtedly one of the best-known 3-D printing designers on the market. This young Dutch woman quickly turned to 3-D technologies, convinced of the benefits they can bring to the fashion sector. In particular, she designed various collections presented at numerous Fashion Weeks. For example, "Ludi Naturae" is a nature-focused collection that included the Foliage dress made up of 3-D-printed leaf-shaped patterns. Using a PolyJet printer, Iris van Herpen was able to create 0.8 mm thick sheets of tulle. She is also behind several 3-D-printed shoes, made in collaboration with the United Nude brand (Smelik, 2020).

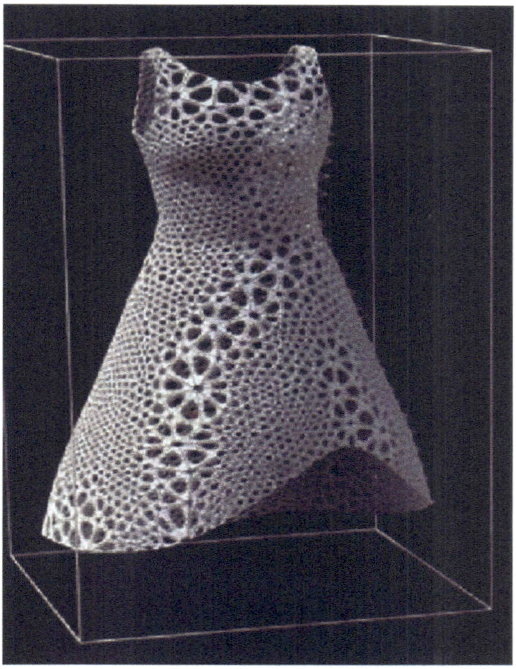

Fig: Danit Peleg 3D Design

Prototyping: production of molds for complex and technical parts. Brands like Cartier or Goossens use 3-D printers to save time and reduce costs.

Since 2011, the Gemmyo brand has used 3-D equipment to produce their simplest parts quickly and reduce storage costs. At first, all the jewelry sold on their website was computer-generated images of 3-D modeled jewelry, of which customers could make their choice. Then, upon request, the jewels were 3-D printed. Today, the firm mainly sells already produced jewelry, in its physical stores and on its website.

10.18 Met Gala 2019, the presence of 3-D technologies

The Met Gala in New York is one of the biggest fashion events of the year. 2019 there was a 3-D printed dress that was developed by designer Zac Posen in collaboration with GE Additive and Protolabs. The dress was named "Rose" and was worn by the model Jourdan Dunn. As the name implies, it was a dress made from 3-D-printed rose petals. SLA technology was used, whereby the printed parts were made with automotive paint to create a special look. Additionally, the designer printed a custom bustier using the same process, which the actress Nina Dobrev wore. Additionally, Zac Posen showed, with various 3-D-printed accessories, how diverse the possibilities of 3-D printing are in fashion (Srinivasan et al., 2020).

10.19 Why: importance and future

Twenty years ago 3-D printing technology was still in its infancy, and had had any major impacts. However, now, the future of 3-D printers is envisaged to bring people endless possibilities, with artists able to ignore the traditional concepts, and show better their creative inspiration. 3-D printing technology will create unlimited value for the apparel and fashion industry in the future. Reports from British media claimed that by 2016, the global USD 13.7 Billion in 2020, 63.46 Billion by 2026.

Unconstrained designers and keen merchants have already begun to use with this technology. In March of 2013, NIKE released a football boot with 3-D printing technology. The use of this new technology means that Nike can make changes based on the boot prototype in a shorter time. This allows design updates that would have taken months to be completed in just a few days. Also, this boot is lighter than the previous design and can increase the user's speed (McCormick et al., 2020).

In fact, in the fashion industry, designers of shoes and jewelry have absorbed this new technology faster. "They are closer to product design and architecture, and the use of 3-D technology in these two fields has some history", pointed out Philip Delamore, director of the London School of Fashion Digital Studio". Especially, shoe designers, are very familiar with 3-D printing technology, and are very excited because it means they can provide something truly unique—and they don't have to buy ready-made heels (models).

No wonder that fashion designers are obsessed with 3-D technology, but of course, this technology also involves some obstacles that need to be overcome. Fashion designers pay attention to the material, they feel the material, see how it hangs and how it swings with the human body's movement. 3-D technology is designed for engineers, not designers; first coming from the automotive and aerospace industries. The development of these materials was also for the application of engineers and there is also a problem of the limited color tones of the materials.

More recently, the designer Iris van Herpen 3-D printed ornaments for one of the dresses presented at Paris Fashion Week 2018. Herpen used the PolyJet technique to create good-quality and smooth-surfaced pieces of resin. Van Herpen, a famous Dutch fashion designer who claimed to be one of the last to use the Internet, has now become a model of the fashion industry's use of 3-D technology. She has created incredible designs that have caused Lady GaGa's stylist to proactively approach her (Hemmings, 2020).

The news about 3D printing in the watch circle is also endless. Last year, in the article "What's more than a watch in Basel" published in "Fashion Times" magazine, 3D equipment and sample printing were shown; Patek Philippe launched the most complicated watch this year. The follow-up model 6002 is also a case model made with 3D printing technology. According to editor-in-chief Pan, the Cartier factory also uses 3D printing applications.

The attraction to the use of 3-D technology is natural. "In the past, I had to have a three-dimensional imagination in my mind first, but to turn it into a real object; I needed to convert it into a two-dimensional look, such as a drawing" claimed Van Herpen, "This way of working, it is too old-fashioned. With the help of 3-D printing technology, I can immediately convert the 3-D picture in my mind into a 3-D model in the computer, and then print it out." 3-D printing technology also allows Van Herpen to no longer be subject to traditional manufacturing techniques and to achieve more complex designs. Iris Van Herpen used Stratasys 3-D printer to design and manufacture basic boots inspired by tree roots, complex structures and full of geometric impact, and brought them to the stage of a recent Paris Fashion Week.

In the fashion industry, which is increasingly pursuing individualization and customization, 3-D technology makes it easier for designers to meet these consumer needs. They only need to make a small change to the 3-D model on the computer; and even consumers can design their own products. From this perspective, as a tool, 3-D is a plus for the fashion industry.

Also, selling 3-D-printed sunglasses is not uncommon, with some companies planning to use 3-D technology to allow consumers to wear self-designed sunglasses. 3-D technology can easily customize products, which makes it an ideal tool for the fashion industry, because, in the fashion world, everyone has a unique body shape and different tastes. However, his 3-D printing technology still needs to overcome some obstacles to be widely accepted by the fashion industry (Propelli & Desai, 2019).

"The biggest obstacle is that people do not use computers in the fashion industry" Janne Kyttanen, a designer at Creative Freedom, teased. This is obviously aimed at designers who are accustomed to drawing on paper, and there is also a "generation gap" problem. "Five years ago, most of our students had no idea about 3-D technology," Delamore said, "but now they see it as another tool." However, there are bold designers who try to combine it with high fashion. In the Van - front charge of the roots of this shoe was born, she was by Stratasys in 3-D to create a women's series printers, shook the entire fashion industry. "When a technology only needs to scan the body, it can make clothes that are a better fit than the tailor's hand. There are no seams. One can no longer say that it has nothing to do with the fashion industry" one designer claimed (Black, 2009).

One industry insider also said: "I am not optimistic about the application of this technology to high-end fashion. Consumers who enjoy high-level customization still hope that traditional tailors will make it meticulously" (Sun & Zhao, 2017). However, in Delamore's view, consumers are not ready to wear printed clothes: "Designers are using 3-D printing technology to create iconic things, but it is not yet a consumer-level technology. We are used to wearing natural materials, even if we wear nylon, but the type, texture, and quality of the folds, what we expect is these, 3-D technology still has a long way to go."

Opponents say that in the future, 3D technology can only penetrate parts of the fashion industry and will never become high-end fashion because it cannot directly replace part of this niche market. In the luxury world at the top of the pyramid, in addition to the top high-end customization, almost all other product lines of the world's top luxury brands use machine assembly lines. The handcrafted quality of haute couture and the origin spirit of luxury have become the only remaining cluster of luxury brands.

References

Black, S. (2009). Editorial in "fashion practice: Design, creative process and the fashion industry". *Fashion Practice: The Journal of Design, Creative Process and the Fashion Industry, 1*(1), 5–8.

Hemmings, J. (2020). Iris van Herpen: Transforming fashion. *Fashion Theory, 24*(2), 287–291.

Kim, H. S., & Kang, I. (2015). Study on status of utilizing 3D printing in the fashion field. *Journal of the Korea Fashion and Costume Design Association, 17*(2), 125–143.

McCormick, H., Zhang, R., Boardman, R., Jones, C., & Henninger, C. E. (2020). 3D-Printing in the fashion industry: A fad or the future?. In *Technology-driven sustainability* (pp. 137–154). Cham: Palgrave Macmillan.

Mpofu, T. P., Mawere, C., & Mukosera, M. (2014). *The impact and application of 3D printing technology*.

Sedhom, R. V. (2015). 3D Printing and its effect on the fashion industry: It's more than just about intellectual property. *Santa Clara Law Reviews, 55*, 865.

Shahrubudin, N., Lee, T. C., & Ramlan, R. (2019). An overview of 3D printing technology: Technological, materials, and applications. *Procedia Manufacturing, 35*, 1286–1296.

Smelik, A. (2020). Fractal folds: The Posthuman fashion of Iris van Herpen. *Fashion Theory*, 1–22.

Srinivasan, R., Mian, A., Gockel, J., & Luehrmann, L. M. (2020). Additive manufacturing and how 3D printing is fighting COVID-19. *Additive Manufacturing, 5*, 1–2020.

Sun, L., & Zhao, L. (2017). Envisioning the era of 3D printing: A conceptual model for the fashion industry. *Fashion and Textiles, 4*(1), 25.

Valtas, A., & Sun, D. (2016). 3D printing for garments production: An exploratory study. *Journal of Fashion Technology & Textile Engineering, 4*(3), 1–4.

Yan, Q., Dong, H., Su, J., Han, J., Song, B., Wei, Q., & Shi, Y. (2018). A review of 3D printing technology for medical applications. *Engineering, 4*(5), 729–742.

Further reading

Noorani, R. (2017). *3D printing: Technology, applications, and selection*. CRC Press.

Parupelli, S. K., & Desai, S. (2019). A comprehensive review of additive manufacturing (3d printing): Processes, applications and future potential. *American Journal of Engineering and Applied Science, 16*, 244–272.

Spahiu, T., Grimmelsmann, N., Ehrmann, A., Shehi, E., & Piperi, E. (November 2016). On the possible use of 3D printing for clothing and shoe manufacture. In *Proceedings of the 7th International Conference of textile, Tirana, Albania* (pp. 10–11).

CHAPTER 11

3-D printing in the food industry

Contents

11.1 What

The application of 3-D printing in the food industry, especially in the past few years, has been related to many start-up companies that have released their own-brand 3-D printers, using various food substances as printing materials (e.g., chocolate, sugar, cocoa powder, dough, raw materials from minced meat, plant-derived hydrogel, sauce, cream). The company's philosophy and concepts have certain market positioning characteristics. One of the modern application directions of industry development is 3-D printing. Since the first 3-D technology was developed (1980), printing methods and other technologies have made revolutionary progress: improving the printing quality of various source materials; significantly reducing the cost of the obtained products and printing equipment; and expanding the subject areas (medical, aerospace, food industry, etc).

3D Printing
ISBN 978-0-323-99463-7
https://doi.org/10.1016/B978-0-323-99463-7.00011-6

11.2 How

Recent models indicate that this process begins with an idea as the main driving factor that motivates a person to carry out a certain activity. Then, considering the existing knowledge, hardware, software, and raw materials, as well as the requirements for the received scientific and technological products, this idea began to be reflected in reality.

In order to realize this idea, a certain knowledge of food chemistry, food hygiene, food production process and equipment, and information technology is required. The necessary machines and equipment include 3-D scanners (if necessary, to create exact copies of the original physical objects of the prototype); personal computers; 3-D printers and other technical equipment designed to prepare the base and bring objects into cooking preparation and canning; 3-D CAD software in the form of programs; raw materials, taking into account the physical characteristics and types of design objects, 3-D printing methods, and further technical processing.

11.3 Why

The requirements for the final product are formed in the creative stage and can correspond to fashion trends, such as the following.

Caramel can be used to make complex objects with boat shapes, coat of arms, colored cubes made of sugar, and three-dimensional inscriptions made of chocolate; individual aspects can be made for a specific group of people, that is, the necessary food is provided in the created food. The required macro- and micronutrients can be included a this device could carry out the work of preparing and cooking pasta, pizza, and other baked goods; and in the future there could be a universal machine to make any food at the touch of a button.

The result of 3-D printing in the food industry will be the food itself, and it will simplify the manufacturing and delivery of products in the distribution system, revise the food industry technology, develop and computerize technical processes, and ultimately improve the quality of human life (Fig. 11.1).

Food 3-D printing is one of the applications of 3-D printing, and it is a rapidly evolving technology. Therefore, new 3-D printers can now be used for various types of food materials. Paste products can to use 3-D printing technology to form any shape or structure, for example, chocolate, fruit

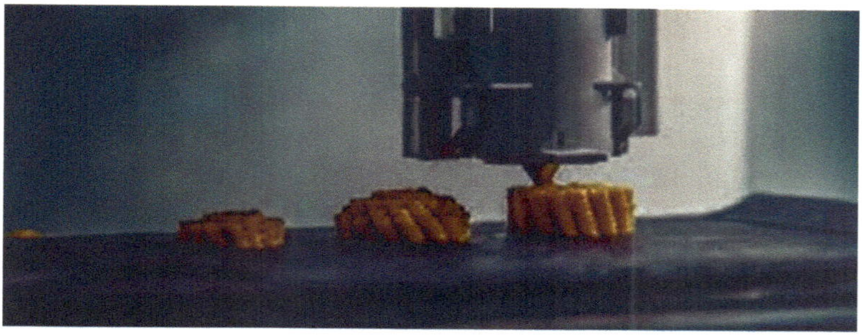

Figure 11.1 3-D printing in the food industry. Pinna et al. (2016).

puree, and pastries. There are currently some projects aimed at objectively and independently evaluating the possibilities and limitations of 3-D printing technology, for example, through actual testing of various food materials. The aim is to develop personalized nutrition according to various dietary needs, such as fortified foods containing vitamin D, calcium, and protein for the elderly, with food being personalized to address specific issues. As an example, it can include missing essential fatty acids and dietary fiber. At present, considering modification and implementation forms, there are eight basic varieties of 3-D printing – about 30; release samples and industrial models – three kinds: FDM, PBP and SLS, which are easy to implement, basically use one or other raw materials details of its possibilities and current and future commercial uses. These technologies are briefly described below (Lipton et al., 2015): (1) FDM (fused deposition modeling), which uses layer-by-layer deposition modeling as developed by S. Scott Trump in the late 1980s and introduced to the market in 1990 by STRATA-SYS. As the original patent expired, the technology became increasingly popular among open-source printer enthusiasts and commercial companies. In turn, the widespread adoption of this technology has led to a significant drop in the price of 3D printers using this production method. The essence of the method is that the printer extrudes the materials (sauce, icing, cheese, dough, chocolate, fruit puree) layer by layer through the dispensing nozzle. Multiple ink cartridges with different materials can be used, and multiple print heads are used separately. (2) PBP (powder binder printing) is powder impact bonding (dropping powder) printing. Z Corporation was first developed by the Massachusetts Institute of Technology in 1993 and obtained an exclusive license in 1995. The essence of the method is that the inkjet print head passes through a layer of powder and selectively applies a liquid binder. Then it applies a thin layer of powder evenly on the entire treated surface and repeats the process. In each layer, the adhered parts of the powder are superimposed on each other. After the printing process is completed, the unbound powder is automatically and/or manually removed, and the remaining powder can be reused. In the initial implementation, starch and sugar were used as powders, and water and food additives [viscosity and surface tension regulators, dyes (for color printing)] were used as binders. In addition, multiple print heads are used to increase the color gamut and enhance the printing process. (3) SLS (selective laser sintering) technology was developed at the University of Texas at Austin in the mid-1980s. In 2001, the printing method was acquired by 3D Systems. The essence of the method includes using a high-power laser to sequentially sinter the powder material layer to provide the partial melting required for the sintered material. Sintering is performed by using one or more lasers to track the contours embedded in the digital model. After the scan is complete, the build platform is lowered and a new layer of material is applied. This process is repeated until a complete model is formed. Before printing, the consumable can be heated to slightly below its melting point to promote sintering. Therefore, it is obvious that due to the particularity of the raw materials used to produce the finished product, the main technology for printing food is through layer-by-layer deposition (FDM) modeling: in most cases, it represents a semiliquid with certain rheological properties. The participation of large food companies in the R&D phase, as well as the true promotion and implementation of this technology in their own production, has formed a competitive market among other players, thereby promoting the development of this scientific direction.

11.4 Who

According to the forecasts, the market size for food 3-D printing will grow at an average annual rate of 16.1% by 2023, reaching US$525.6 million, and then US$1 billion by 2025. However, this is far from the maximum capabilities of the larger food technology sector. By 2022, the global food industry is expected to reach US$250 billion, and 3-D printed food is a very healthy market segment. Data Bridge Market Research predicts that the food 3-D printing market will grow during the forecast period from 2020 to 2027, reaching US$412.3 million by 2027 (Godoi et al., 2016). At the same time, during the aforementioned forecast period, the compound annual growth rate will reach 52.30%. The growing demand for healthier and more personalized foods is driving the development of the micro-encapsulated food market (Fig. 11.2).

11.5 3-D printing products reduce waste

Globally, about one-third of food is wasted. Farmers, wholesalers, supermarkets, and (government) agencies work individually or collectively in local markets to reduce the amount of food they waste. The huge quantity is a huge motivation to find new opportunities to reduce waste. Food 3-D printing could play an interesting role in this area. However, supermarkets are not the only ones throwing away overcooked food. Customers who expect perfect food are also guilty of wasting food, with imperfect but edible vegetables and fruits often discarded.

11.5.1 Printer-food 3-D printing

11.5.1.1 Flow focus

It may not seem significant, tut 3-D printing food with a small printer can achieve many things. Focus 3-D printing is used by professional chefs all over the world, and is controlled by WiFi. Reusable cartridges (syringes) are used to fill edible pastes. Smink, a Michelin-starred restaurant in the Netherlands, uses 3-D printed food to create intricate dessert patterns. The current price of ByFlow is about US£4300 (3900 Euros) (Pallottino et al., 2016).

11.5.1.2 MM use

The MMuse-Chocolate 3-D printer is unique, it can process chocolate chips up to 4 mm in diameter and melt them for extrusion. It is very strong

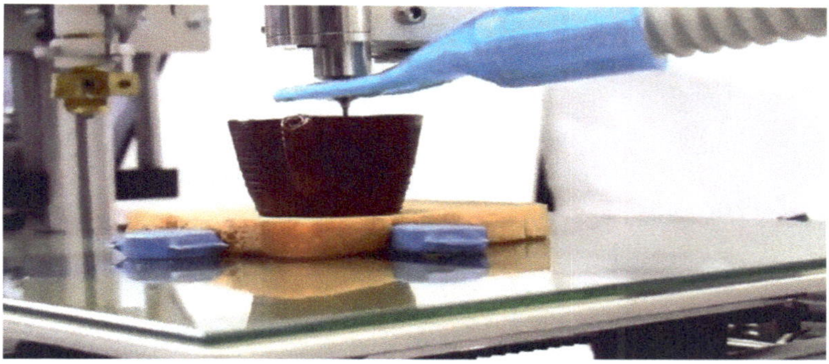

Figure 11.2 Personalized foods by 3D printers. Sun et al. (2015).

and weighs approximately 25 kg (approximately 55 pounds). Although desktop computers can be used, 3-D printing food on it is expensive. The printer is manufactured in China and is priced at approximately US$5700 (Pallottino et al., 2016).

11.5.1.3 Fodini

This is one of the most popular food 3-D printers of all time, and rightly so. Natural Machines Foodini 3-D-printers can use almost any ingredient, from jelly to delicious burgers. Using five different cartridges at the same time, Natural Machines enables a healthy diet. Therefore, they provide new recipes for beginners before they are able to create their own recipes. The cost of the Foodini is about US$4000 (Pallottino et al., 2016).

11.5.1.4 Home 3-D printer

There are also food-grade 3-D printers for the consumer and semi-professional markets. 3-D printed food is not limited to special events and gourmet restaurants, with several machines specially designed for home hobby users. However, they are not very popular.

11.5.1.5 Fungus

Mycusini is a chocolate 3-D printer sold by Print2Taste in Germany. It has an excellent design and is compact enough to be used on a kitchen countertop. Its built-in battery allows it to print offline for up to 2 hours. Its stainless steel cartridges are easy to fill and clean. However, refills must be purchased directly from Print2Taste. The price of Mycusini is approximately US$370 (~339 Euro) (Pallottino et al., 2016).

11.5.1.6 Pancake robot

Calling the PancakeBot a food-grade 3D printer might be a bit inaccurate, as it only makes pancakes. But anyway, it is a really cool robot. The latest version of PancakeBot v2.0 has many control overprint settings as well as a simpler user interface and a heating surface to cook the dough. PancakeBot 2.0 starts at $300 (Pallottino et al., 2016).

11.5.1.7 Prucusini

Procusini 4.0 is another food 3-D printer from Print2Taste, which is aimed at the semiprofessional market. The cost of this unit is approximately US$2600 (~€2382). During the printing process, the ink cartridge can be heated to 60°C to facilitate the extrusion of thicker materials (Pallottino et al., 2016), for example, marzipan or fondant that can be used to decorate cakes. Procusini is advertised as a plug-and-play system, and the refills must be purchased directly from the manufacturer.

11.5.2 Do 3-D food printers they really have the potential to conquer the kitchen?

On the other hand, one of the greatest difficulties with this type of system may be in providing them with more powerful features. Could we really make hamburgers, pasta, or steaks, or complex dishes with the flavor, texture, and appearance obtained by traditional methods?

In my opinion, it is hard to imagine that these chefs would allow it in short term. However, the technology may gradually encounter increasingly ambitious challenges. However, technology may meet increasingly ambitious challenges gradually. It may soon be possible to cook as you go, and some dishes can be prepared as a whole. We may soon also see the development of new dishes and shapes that will replace traditional shapes much like ice cubes from 200 years ago. It is impossible to understand the cream industry in the 21st century because the concept does not exist and there is no 3D printer market for ice cream, chewing gum or carbonated drinks.

To support the implementation of these 3D printing systems, there are also issues that consumers do not consider today because they have many operating challenges, but there are also many advantages, therefore, 3D printing will be very important in the future.

For example, a globally interconnected system where food production is nonlocalized, where the diet can be personalized to adapt to everyone's needs, and the same machine is responsible for their planning and preparation, will give us more free time and eliminate many daily tasks.

Being able to choose our favorite food, taking into account our specific nutritional needs, our daily activities, and our budget, the system will prepare the food required for immediate consumption. This is a dream that we may realize in the next few years.

On the other hand, due to population growth, natural resources are increasingly scarce. The Earth's resources are limited, and the methods of obtaining food in the future may be very different from today. The current breeding of cattle, sheep, birds, etc. is unsustainable and will not be able to meet the exponentially growing population demand over the long term. Thus there must be other sources of protein that are easier to produce and more effective, including sources that we reject today, or sources whose appearance, aroma, taste, and texture are unsatisfactory. Therefore, it will necessary for the large-scale use of algae and insects, and the use of production systems based on cell tissue. In this case, transportation of raw materials for preparation, subsequent texture, flavoring, and assembly at the consumer site may be the most suitable solution. The 3-D printing and cooking system will have a bright future in this area (Nachal et al., 2019).

Similarly, from the perspective of the sustainability of our society, reducing carbon footprint and energy consumption will play an important role in the design of new production systems—including the transportation of basic ingredients such as protein, carbohydrates, fatty acids, vitamins, etc. In addition, flavors and flavoring agents would be more effective for transporting than the entire product as produced today, as 70%—90% of the weight of food is water. On the other hand, today's food waste is huge, and a nonlocalized production system may mean a significant reduction in such waste.

Another great advantage of this system is that it can adapt to individuals' dietary needs. Today, nutritional deficiencies often occur in populations of both developed and developing countries. The incredible paradox is that in the developed world, nutrition is causing serious obesity and a variety of related diseases, while in developing countries, problems caused by malnutrition often appear due to a lack of certain essential nutrients.

In both cases, having a low-cost, flexible, and individualized national food system would make it possible to quickly utilize all available sources of nutrition at a reasonable price.

Let us assume that the basis of a country's diet is rice, but the contributions of some essential vitamins and amino acids are very limited. 3-D food printing systems will allow the use of local raw materials, obtained in small batches and supplemented by the addition of minerals, vitamins or other sources of protein to make up for nutritional deficiencies, resulting in balanced dishes with acceptable texture and flavor. Or take advantage of its technologically modified unusual sources.

11.6 Conclusion

The weakness of this technology is the scarcity and relatively high cost of 3-D printing equipment and raw materials. Therefore, the cost of a 3-D printer with the least features starts at US$500. In the United States, a more professional model starts at $1200, therefore, considering the cost of raw materials and services, not every catering company can afford one. It should be noted that these relatively inexpensive models mainly use sugar as a raw material, which greatly limits the range of nutritional value of the product (Pitayachaval et al., 2018). Another requirement of the 3-D printing food technology process is the technician's knowledge of the specialized software used to build 3-D models, despite the fact that modern 3-D printers have a considerable standard form basis embedded in their memory. At the same time, these weaknesses provide opportunities for food obtained through 3-D printing methods, which are associated with expanding production and reducing costs, which will lead to partial replacement of traditional cooking methods in the future. In addition, considering that foods printed by 3-D printers are highly personalized, their scope can be expanded by adding food additives and functional national food ingredients, which will allow for the production of fortified foods. Expanding the range of raw materials and increasing the number of print heads in 3-D printers will make it possible to produce products with specific biological and energy values, and physical and sensory characteristics, that is, specialized food. The use of 3-D printing technology could solve the problem of the lack of biologically active substances in the diets of people with allergies.

References

Godoi, F. C., Prakash, S., & Bhandari, B. R. (2016). 3d printing technologies applied for food design: Status and prospects. *Journal of Food Engineering, 179*, 44–54.

Lipton, J. I., Cutler, M., Nigl, F., Cohen, D., & Lipson, H. (2015). Additive manufacturing for the food industry. *Trends in Food Science & Technology, 43*(1), 114–123.

Nachal, N., Moses, J. A., Karthik, P., & Anandharamakrishnan, C. (2019). Applications of 3D Printing in food processing. *Food Engineering Reviews, 11*(3), 123–141.

Pallottino, F., Hakola, L., Costa, C., Antonucci, F., Figorilli, S., Seisto, A., & Menesatti, P. (2016). Printing on food or food printing: A review. *Food and Bioprocess Technology, 9*(5), 725–733.

Pinna, C., Ramundo, L., Sisca, F. G., Angioletti, C. M., Taisch, M., & Terzi, S. (2016). Additive manufacturing applications within food industry: An actual overview and future opportunities. In *21st summer school francesco turco 2016* (pp. 18–24). AIDI-Italian Association of Industrial Operations Professors.

Pitayachaval, P., Sanklong, N., & Thongrak, A. (2018). A review of 3D food printing technology. In *MATEC web of conferences* (Vol. 213, p. 01012). EDP Sciences.

Sun, J., Zhou, W., Huang, D., Fuh, J. Y., & Hong, G. S. (2015). An overview of 3D printing technologies for food fabrication. *Food and Bioprocess Technology, 8*(8), 1605–1615.

CHAPTER 12

3-D printing in the footwear industry

Contents

12.1 What

Over the past decade, additive manufacturing or 3-D printing has affected many different industries. One of these is the fashion and footwear industry. The continuous advancement of 3-D printing technology has promoted the development of new solutions, thereby changing the fashion and footwear industries. Although it is too early for 100% 3-D printing of shoes, some global shoe giants have integrated 3-D printers and additive manufacturing into their production systems and started mass production. With the addition of innovative 3-D printing vendors such as Carbon and Form labs, brands such as Adidas and New Balance are becoming increasingly successful in 3-D printing shoes. Not only the world's giant shoe manufacturers, but also individual design companies can use additive manufacturing to produce uniquely designed 3-D-printed shoes. 3-D printing is a technology that clearly is expanding and is not new. Although it is apparent that there is still a lot to do in this industry, there are

3D Printing
ISBN 978-0-323-99463-7
https://doi.org/10.1016/B978-0-323-99463-7.00012-8

already many companies in the footwear industry that are taking advantage of the huge possibilities offered by 3-D printing. According to available data, in the next 10 years, 3-D printing in the footwear industry is expected to bring about $6.3 billion in revenue to companies that use the technology in their production processes (Vanderploeg et al., 2017).

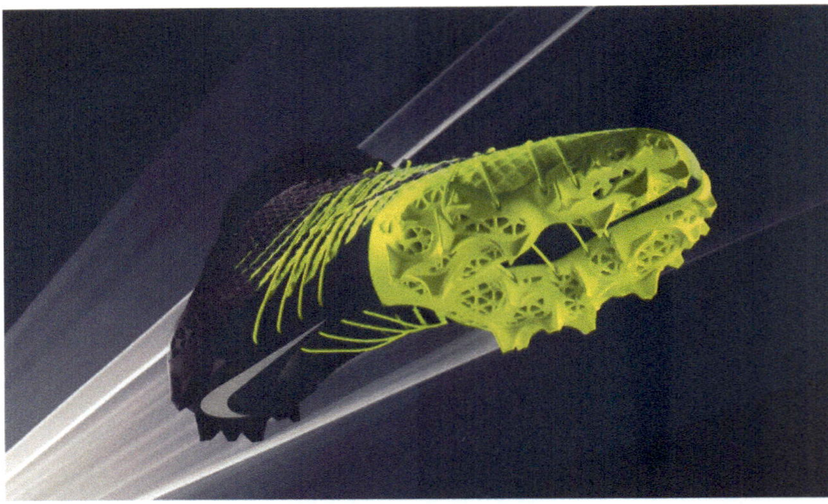

(From Source: Nike football accelerates innovation with 3D printed "concept cleat" for shuttle. (2021). Retrieved 24 February 2021, from https://news.nike.com/news/nike-football-accelerates-innovation-with-3d-printed-concept-cleat-for-shuttle.)

12.2 When

The use of additive technology or 3-D printing in today's footwear products is not only a fashion trend, but also a reasonable modern production method. In the manufacturing of professional sports footwear (soccer shoes, running shoes), 3-D printing allows for a reduction in technical costs and implementation of the most complex ideas to reduce weight, increase comfort, and improve product quality. In addition, volume printing is essential for the styling and production of exclusive designer styles.

12.3 Where

In order to manufacture the sole of a model on a 3-D printer, a flexible, soft plastic material is used. This is a plastic thread that has great resistance under pressure. Young designers also believe that other plastics could be used for 3-D printed shoes. This is a nylon powder that has been processed by selective laser sintering (SLS) to quickly melt and solidify shoe soles. The designer uses shrink yarn to make the upper part of the shoe. Here, a software 3-D model is used. Currently, new materials for making shoes on 3-D printers are being developed. Recreus has recently developed and is already producing Filaflex flexible filaments. This material is different from the traditional ABS and PLA plastics used for 3-D printing and will not harm human skin when wearing 3-D printed shoes. On the contrary, Filaflex shoes fit the feet perfectly and are very comfortable even for long periods of wear.

Using personal 3-D printers, simple products can be printed, such as shoe soles, slippers, or the decorative top layer of finished shoes. To do this, 3-D models of future products, special software, elastic wear-resistant consumables, and 3-D printers are needed. At home, it is entirely possible to use FDM technology to print shoes on a traditional printer.

12.4 Who

Large companies in the sports shoe industry have begun to attempt to take advantage of this huge opportunity, with the five major companies in the field, Adidas, Nike, Under Armour, New Balance, and Reebok, carrying out various projects based on 3-D printing, which allows them to gain experience of all the possibilities of 3-D printing to meet the demands of global trends, mass customization, and mass production (Amza et al., 2019).

12.4.1 Adidas, 3-D-printed midsole

Adidas is one of the latest companies to incorporate 3-D printing technology into its production process. Nevertheless, they decided to make up for this late take up by betting heavily on this technology. The company has produced more than 100,000 pairs of shoes, including midsoles, produced using the technology based on the carbon digital light synthesis system, which can be quickly printed using layerless photopolymerization. The next step is to use a special type of durable and flexible liquid resin that combines thermosetting polyurethane and a special light curing agent to print the complete shoe directly on a 3-D carbon printer.

12.4.2 Nike, custom football boots

Nike is a pioneer in the extensive use of 3-D printing to develop high-performance products. Its first major project was the manufacture of ultra-light football shoes: Nike Vapor Laser Talon. In this project, 3-D printing allows testing, iteration, and creation of shapes that cannot be achieved by traditional manufacturing processes, helping the company to push the boundaries of innovation faster. Nike is currently experimenting with the production of final products, especially its Nike Flyprint project, which is the first 3-D-printed textile upper for high-performance footwear. In this case, the technique used is solid deposition modeling, which is a process of unrolling, melting, and layering TPU filaments from a spool.

12.4.3 New Balance limited edition sneakers (Amza et al., 2019)

Like Nike, New Balance is also one of the pioneers in the use of 3-D printing in its shoe manufacturing. In 2016, it started manufacturing Zante Generate sneakers, the first 3-D-printed sneakers that can be purchased commercially. However, there are only 44 pairs in the brand's online and Boston stores. New Balance recently introduced Russell Finex's automated finishing technology in its additive manufacturing laboratory, which shows that the company is working hard to achieve more use of 3-D printing as a means of automating the midsole manufacturing process.

12.4.4 Under Armour is marked by innovation

Sportswear brand Under Armour regards innovation as one of the key elements of its business strategy and it is one of the first companies to adopt additive manufacturing in the production of its end-use parts. Therefore, in 2016, they launched the UA Architect series that includes the first ever 3-D-printed midsoles. The midsole printed in 2017 introduced a new TPU material developed by Lehmann & Voss & Co.

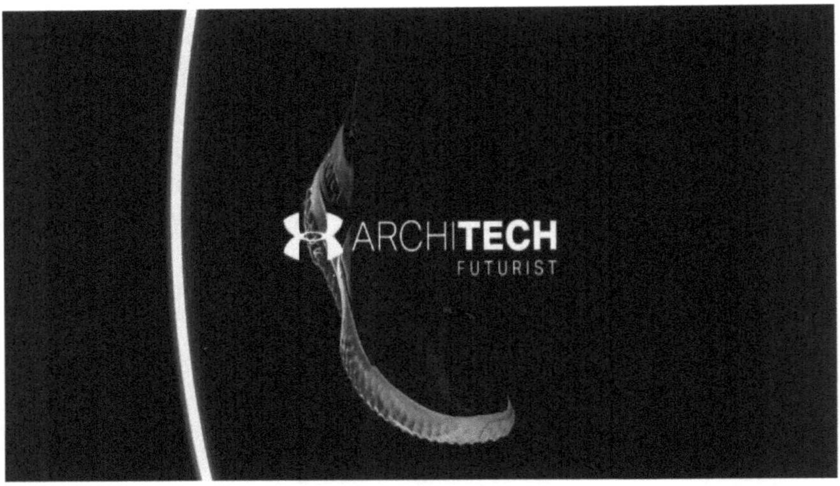

(From Source: Han, G. (2021). The UA ArchiTech futurist floats atop 3-D printed midsoles. Retrieved 24 February 2021, from https://design-milk.com/ua-architech-futurist-3-D-printed/.)

Under Armour is significantly increasing the production of shoes with 3-D-printed components. Therefore, although the original Architect series only had 96 pairs, it is expected to reach more than 2000 pairs.

12.4.5 Reebok, 3-D-printed shoe upper

Reebok later applied 3-D printing to the field of sports shoes, and in 2018 launched the Liquid Floatride Run shoes, which introduced 3-D drawing technology and then printed using liquid elements such as "Flexweave" materials. This creates a shoe that is 20% lighter than the original version and also includes new features such as liquid shoelaces that are 3-D printed directly on the back of the shoe (Amza et al., 2019).

12.5 Why

Footwear 3-D printing is famous because it allows for personalization, but most importantly, it is an unparalleled tailor-made product. In addition, this is an ever-evolving industry, and the materials and processes used are constantly improving. 3-D printers are also becoming increasingly efficient. This leaves room for unlimited creative possibilities. This innovation has just started because there is no 100% 3-D finished product yet. On the consumer side, the possibility of creating tailor-made and customizable shoes is very attractive. In addition, for people who have difficulty wearing shoes, additive manufacturing will help solve various health-related problems in the sole or the shoe itself.

In fact, additive manufacturing greatly reduces waste in the product manufacturing process. However, due to the lack of flexible materials, there is currently no material to make 100% 3-D shoes. Some start-ups also ask customers to send their old 3-D shoes back, melt them, and then use them to make a new pair of shoes. Therefore, this aspect of additive manufacturing plays an important role in helping brands and consumers to respect the environment.

The whole sneaker is made of one piece, so it can be fit perfectly, without tearing the seams, or collapsing in the weakest place after too much moisture enters. With the help of special robots, a viscous liquid is squeezed onto a flat surface, forming something similar to a flexible rubber structure. Then it blends into the shoe, from the outsole to the laces, forming a flexible outer cage, which moves with the shoe and reshapes it as needed. The resulting ideal shape makes the shoe fit like a glove. If a person has individual ankle structure characteristics, the problem can be eliminated by making the necessary changes to the original model. The most amazing thing is that 3-D printing technology allows the creation of individual parts when manufacturing shoes, or they can be "print" whole. 3-D printing in the production of footwear reduces the weight of shoes. Due to the complex internal structure of the printing, there are more opportunities to change the elasticity and stiffness. Also, the broadest field of designing and inventing new shapes that cannot be created using traditional methods (Spahiu et al., 2016).

12.6 How

According to extensive research, photopolymerization is the company's most commonly used additive manufacturing technology for 3-D shoe design. It is very simple, because it has the highest ratio between printing speed and accuracy, so it can be used to produce sufficiently large quantities while maintaining the quality of 3-D printing.

In footwear manufacturing, 3-D technology is used in different stages. There is a need to scan for measurement and accurately create a digital model, use three-dimensional design in the design, and use printing when making the finished product. Professional 3-D equipment can be provided for performing any stage.

12.6.1 Use of 3-D printing variants in footwear production

The main application areas of additive technology in the footwear industry are specialized orthopedic models, sports, and design styles. Leading brands such as Nike, Adidas, Reebok, and New Balance use 3-D printing in the mass production of sports shoes. At the same time, 3-D printers enable:

- The prototype of a new model to be made;
- Finished sports shoes, with no need to connect individual elements, and no need to use glue and seams;
- Personal orthopedic insoles for sports, medical, and protective footwear;
- Individual parts—soles made of thermoplastic polyurethane, uniquely shaped heels, protective linings, and decorative elements;
- Unusual designer shoes for fashion shows, theater performances, and future movies, can be manufactured.

Through rapid prototyping, many options for future shoes are printed out to evaluate their ergonomics, comfort, and whether they meet the expectations of the target audience. The high printing speed allows visualization of any design phase of the evaluation and testing of intermediate samples. The actual production of shoes on 3-D printers is not yet very common. However, this is a very promising direction. Today, leading brands are printing unique lightweight sole designs that cannot be created by traditional molding. Advances in technology and the emergence of new consumables have made 3-D printing of shoes more affordable.

Extensive research has supported the design of Futurecraft 4D. The heel is designed to absorb impact, the middle part facilitates the transition from the heel to the toe, and the toe part is designed for takeoff. This is just another goal of footwear: in addition to cushioning, it also provides "energy feedback," which is a technical term for rebound. Each of the 20,000 individual points distributed throughout the sole fulfills these functions (Spahiu et al., 2016).

12.6.2 The future of 3-D printing in the footwear industry

Although the sports footwear industry seems to focus more on fashion than sports, innovation is still emerging, allowing an increasingly realistic future to be imagined. Just as nylon and Velcro made great progress at the time, today, 3-D printed can be considered as a novel method of production. 3-D printing aims to become the next solution for footwear customization. Through tread research and foot shape knowledge, it is possible to produce customized shoes that are completely designed for a specific foot. The idea of scanning your feet and printing a 100% customized sole based on your body type, your steps, your weight, and the activities you will be doing is not far away. This will be the next major change in the footwear industry in the next 5–10 years. In 2016, at the Boston Marathon, New Balance launched a shoe designed and manufactured in its Lawrence factory, manufactured using 3-D printing technology. The price of the Zante Generate was US$400, but only 44 pairs were produced (Spahiu et al., 2019).

Although the foam midsole will flatten and lose some shape over time, all signs indicate that the 3-D molded midsole solves this problem. Carbon technology solves the other major shortcomings of the typical 3-D printing process, such as time and price: the construction time of Futurecraft 4D is no more than 1 hour. While the first batch will only be 5,000 pairs, Adidas aims to produce 1 million pairs in 2022 and gradually increase it. The cost of $300 does not seem to be overly expensive, especially it is considered that it is the first version of a product that is expected to change the well-known sports footwear industry. As there have been enough prototypes, it is time to produce them on a larger scale (Weller et al., 2015).

References

Amza, C., Zapciu, A., & Popescu, D. (2019). 3D-Printed shoe last for bespoke shoe manufacturing. In *MATEC web of conferences* (Vol. 290, p. 04001). EDP Sciences.

Han, G. (2021). *The UA ArchiTech futurist floats atop 3D printed midsoles.* Retrieved 24 February 2021, from https://design-milk.com/ua-architech-futurist-3d-printed/.

Nike football accelerates innovation with 3D printed "concept cleat" for shuttle.(2021). Retrieved 24 February 2021, from https://news.nike.com/news/nike-football-accelerates-innovation-with-3d-printed-concept-cleat-for-shuttle.

Spahiu, T., Grimmelsmann, N., Ehrmann, A., Shehi, E., & Piperi, E. (November 2016). On the possible use of 3D printing for clothing and shoe manufacture. In *Proceedings of the 7th international conference of textile* (pp. 10–11). Tirana, Albania.

Spahiu, T., Piperi, E., Ehrmann, A., Almeida, H. A., Ascenso, R. M., & Vitorino, L. C. (October 2019). 3D printing: An innovative technology for customised shoe manufacturing. In *International conference of progress in digital and physical manufacturing* (pp. 171–180). Cham: Springer.

Vanderploeg, A., Lee, S. E., & Mamp, M. (2017). The application of 3D printing technology in the fashion industry. *International Journal of Fashion Design, Technology and Education, 10*(2), 170–179.

Weller, C., Kleer, R., & Piller, F. T. (2015). Economic implications of 3D printing: Market structure models in light of additive manufacturing revisited. *International Journal of Production Economics, 164*, 43–56.

CHAPTER 13

3-D printing in healthcare

Contents

13.1 How

3-D printing is becoming increasingly popular in industries that conform to the 4.0 concept because it allows products to be personalized to meet specific customer needs and provides greater intelligence and agility to the production process. However, it is necessary to pay attention to the care that 3-D printers need in this situation (Hornick, 2017). It is important to realize the positive impact of technological innovation on society every day. The presence of technology in health care is becoming increasingly important, bringing benefits such as faster diagnosis and modern disease treatment. The data show a close relationship between technology and health: about 9000 medical applications are available to Internet users

3D Printing
ISBN 978-0-323-99463-7
https://doi.org/10.1016/B978-0-323-99463-7.00013-X

worldwide, and 72% of Internet users seek information on medical issues on the Internet. Integrating 3-D printing technology into medicine is a huge challenge. 3-D printing technology is of great assistance to medical professionals and patients, and ultimately makes assistance costs lower and services better. Worldwide, this technology is still used to build prototypes in most cases, thereby increasing the flexibility of the production line. Small-volume spare parts manufacturing has always been a common goal for 3-D printers, but in the industry, 3-D printing can bring about many other benefits, such as:

- Design improvements;
- Reduced tool costs;
- Significantly reduced industrial;
- Speeded up product development;
- Cheaper production.

13.2 What

Like any machine, 3-D printers based on additive manufacturing require special attention to maintain their high performance and not hinder factory production due to failures. Therefore, continuous maintenance is one of the top tasks. Like any equipment, printers require regular maintenance to ensure that they can function uninterrupted and reach the expected output. For this reason, it is essential to check the lubrication level, and clean the nozzles, cover gears, and printing surface, with lubrication being carried out monthly.

It is also very important to take care of the environment in which the machine is located, as temperature, humidity, and waste will all affect the printing of 3-D objects. If the printer is exposed to a continuous cold environment, the factory will need to turn on the air-conditioning which will affect the products printed with acrylonitrile butadiene styrene (ABS). On the other hand, a humid environment can make the filaments used for additive manufacturing brittle and contain bubbles. The dust generated by certain manufacturing methods can also damage the printer itself and the object being printed, causing it to deform. Therefore, it is important to avoid these factors.

The filament is an integral part of the object, and so it deserves special attention. No matter what material it is made of, it must always be in good condition. In addition to the above-mentioned precautions, in order to prevent moisture, cold, and dirt ingression, they must be properly stored and packaged to maintain their integrity, thereby ensuring the quality of the product. In addition, it is important to adjust the coil so that these filaments do not break (Eshkalak et al., 2020).

Using the latest firmware updates is also important for the final result. Only in this way can it be ensured that the printer is correctly compatible with the software resources.

With the new coronavirus pandemic, many people have engaged in 3-D printing on the Internet to direct this activity to helping the healthcare sector. 3-D printing in the healthcare field has been used to make preoperative research models and personalized medicines. Therefore, these hobbyists are using this tool to produce PPE for health professionals, such as face protectors and masks, as well as other equipment used by healthcare workers in their daily lives.

The idea has been to use models found on the Internet and improve them with the help of doctors to make the equipment safe and comfortable. In this way, it will not interfere with the daily work of the hospital. However, most of these PPEs have not yet been approved. Therefore, first of all, they need to go through technical and thorough analysis by responsible professionals so that the equipment can be safely used. The main considerations for these products are discussed below (Trenfield et al., 2019).

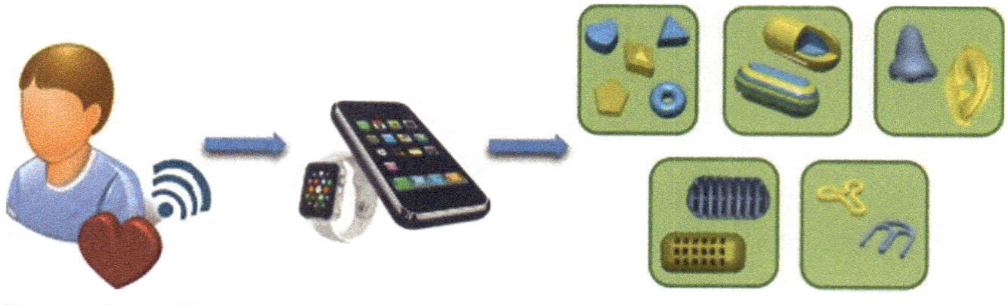

(Source: https://hub.packtpub.com/how-are-mobile-apps-transforming-the-healthcare-industry/.)

13.3 When

The official start to 3-GD printing can be traced back to 1982, when Mr. Chuck Hull invented stereolithography technology and founded 3DSystems. He filed a patent in 1985 and obtained the patent in 1986. The next step was to set up his company and he began to distribute 3-D printers in the 1990s. He laid the foundation and paved the way for all those who have followed, while allowing himself and his company to keep up with the pace of innovation—in fact, his patented concept of creating physical objects as a series of overlapping 2-D layers remains valid today.

In 1986, Carl Deckard, Joe Beaman, and Paul Forderhase (and many other researchers) used selective laser sintering, a similar method, but brought about a change: they replaced the 1986 concept with Chakhel's resin plus nylon, which is a powdery liquid. Because the powder is very strong and hardly needs any modifications, it brings with it many practical benefits.

Mr. Kremp invented fused deposition modeling or 3-D printing of liquid materials in 1988. This puts aside the laser and dust, and uses molten plastic to "spread" the object layer by layer. He later founded Stratasys, which is now an industry leader.

In 1993, it was the turn of the Massachusetts Institute of Technology (the Institute of Technology in Boston). The fourth printing technology was a technology that promotes color printing, called "3-D printing" which can print up to 28 types. In addition to color printing, this is a very valuable but expensive technology for printing very realistic objects.

The selective laser melting process was initiated in Germany at the Fraunhofer Institute in 1995. For the first time in the field of 3-D printing, it was realized that very solid objects can be created, the density of which had been unimaginable in traditional industries. Due to this process, metal powder can be melted and an object with a density of 98% can be obtained.

In 2002, "electron beam melting" or fusion to an electron beam, which a high-energy source, was invented, using a material in the form of "microparticles" composed of appropriately concentrated beams and accelerated electrons, "resulting in its complete merger. Using this method, even metal objects with a density equal to 99.98% can be obtained. Therefore, today, almost everyone can use a 3-D printer for rapid prototyping.

13.4 Where

3-D printing technology is nothing new in the medical field as it was started in the 1980s. Its main application was to manufacture parts for the automotive industry and create product prototypes before authorizing production. Since then, many other types of products have been manufactured through 3-D printing, such as clothing, weapons, and even toys.

3-D printing is a method that allows almost anything to be printed. However, this is not printing on a sheet of paper, but constructing an object entirely in 3-D format. Thanks to this technology, replicas of prostheses and organs have begun to be created. Customized for individuals, they can replace lost body parts, such as the legs, nose, and even bones.

According to the WHO, we are facing the greatest health crisis of our time due to the new coronavirus (COVID-19). Therefore, society is united in a single goal of combating this virus, which is not restricted to any particular region or belief, race or social status. Therefore, many professionals from all walks of life are working hard to combat COVID-19. Companies that use 3-D printers are also part of this fight. The use of 3-D printing in the sanitary field, especially under the current circumstances, is a good choice.

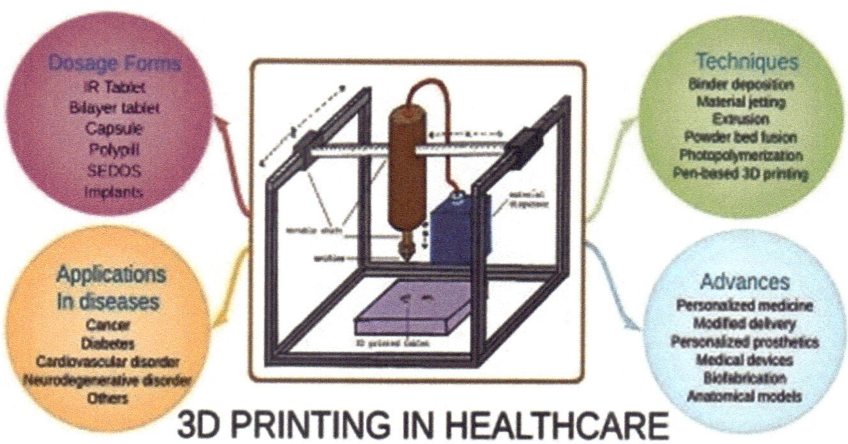

(Source: https://www.sciencedirect.com/science/article/abs/pii/S0753332218318407.)

13.5 How

In order to be able to print, the product design must be prepared with the aid of a computer program or 3-D scanner. After this stage, the product is printed, usually using materials composed of plastic polymers. The device can print 3-D parts layer by layer, such as fingers, skulls, or teeth.

One of the major advantages of this technology is customization. In order to create certain parts of the patient's body, magnetic resonance images or computed tomography scans are used, which guarantee an accurate copy. It is believed that in the future it will be possible to use living cells as raw materials for the construction of body parts, thus eliminating the long waits for transplantation. For this reason, 3-D printing has received increased investment.

13.6 Why is 3-D printing important in medicine?

The prostheses produced by 3-D printing are still new to some parts of the world but will soon become commonplace. Some possible uses are described below.

13.7 Organ transplant

In terms of organ transplantation, it can be very difficult to obtain a compatible donor or deal with the possibility of organ rejection. However, 3-D printers can solve these problems.

The working principle of transplantation is as follows. The stem cells are mixed with fragments of the organ to be reconstructed and deposited in a glass receptacle, which has the shape of the desired tissue. Using the "mold" as a reference, the cells begin the process of multiplying to produce the shape of the desired organ.

Isn't it amazing? Therefore, this is a mechanism that could end the demand for donations forever in the near future.

13.8 Surgical plan

3-D printing helps reduce the risks of surgery. By creating organ replicas, doctors can study all the anatomical features of patients. Therefore, by analyzing the structure from various angles, the details of the lesion to be treated can be identified, and a higher treatment accuracy can be ensured.

13.9 Temperature and sterilization

All items in hospitals must be disinfected, especially in the intensive care unit (ICU). This is usually done by exposing the object to a high temperature; in more specific cases, it can be chemically treated. This procedure is very old and has been carried out in hospitals for many years. However, the most commonly used material in 3-D printing is polylactic acid (PLA) plastic, which is not resistant to high temperatures. If PLA parts are used in a hospital, they may suffer major damage after disinfection.

Common materials used in 3-D printing also do not support chemical processing. Therefore, only a few printed materials can be used for such applications. ULTEM is one of these materials; its printing temperature is 390°C, but some printers used to produce masks have a maximum temperature of 250°C.

13.10 Pointed equipment

As is well known, medical personnel wears clothing from the head to the toes to avoid contact with viruses, etc. This is essential to protect them from becoming ill, and if they get sick, the capacity of the hospital will be affected. Therefore, another precaution that should be taken is the supply of equipment with pointed tips to prevent puncture of protective clothing. Therefore, it is necessary to analyze whether the element is really safe, so as not to cause risks to professionals.

13.11 Function

One of the required pieces of equipment in a hospital is the ventilator, which is extremely important for patients with respiratory problems. There has been a suggestion to use 3-D parts to create these device, which seems to be a viable idea. The alternative has proven to be very interesting; in some projects, the ventilator can serve two or more patients at the same time (Trenfield et al., 2018).

On the other hand, this kind of airflow requires special attention, because parts produced using 3-D FDM printing technology have a porous surface finish, which interferes with the airflow. Therefore, only parts manufactured using 3-D SLA printing technology can be used for these respirators, because these parts have a smoother surface and the materials used are very high temperature resistant.

13.12 Porosity

Since there is no approved equipment, engineers and doctors are required to perform a series of technical analysis to verify the use of PPE produced by 3-D printing enthusiasts for use in health care. Therefore, another precaution that must be taken when printing these devices is the surface finish of the produced parts, which usually leaves open pores and spaces for bacteria, dust, viruses, etc. to accumulate. Therefore, to include 3-D printing in the fight against coronavirus, it is necessary to make equipment that does not pose a risk to the health of the user or patients.

13.13 Prosthetic structure

Prostheses are a foundation of the process of restoring self-esteem. In addition to facilitating reintegration into society, they also enable patients to perform tasks more easily. These prostheses include hands, legs, arms, and other parts of the body.

The cost of traditional prostheses can be prohibitively high. In addition to being cheaper, 3-D printed prostheses can be more life-like and lighter.

13.14 Medical intervention

In order to discover what can be done to help combat the new coronavirus, the many volunteers may end up hindering the important work of healthcare professionals. To solve this situation, we must ensure that the ability to produce such equipment is of sufficient quality so as not to interrupt medical work or cause any harm to the wider society.

13.15 Countless possibilities

Most 3-D printing volunteers have no medical knowledge or support, so they will likely not know everything that could go wrong. Therefore, one way to overcome this problem is to include medical knowledge in discussions and actions to ensure that all known problems are overcome before continuing production. Obviously, people who are committed to using 3-D printing in the health field have the best intentions and their only goal is to provide help at this crucial time to prevent the situation from deteriorating (Aquino et al., 2018).

During the manufacturing process of these PPE, other precautions must be taken, namely:

- Use 70% alcohol to clean the printer surface and working area;
- Wear masks and gloves during the manufacturing process, and store the parts in sealed packaging at the end of the operation;
- When sending the equipment, describe the production environment and conditions;

- After finishing work, clean the parts with soap and water;
- Each part should be packaged separately.

3-D printing in the healthcare sector could be an excellent solution in helping to fight the COVID-19 coronavirus. However, this entire process must be done very carefully to truly help these professionals. It is critical to produce these devices with the utmost care and attention.

13.16 Who

At the Dutch University Medical Center, surgeons successfully replaced part of the skull of a 22-year-old woman with a 3-D model made of plastic. The prosthesis completely replaced the damaged part. However, the young woman's skull thickened over time, which posed a serious threat to her future health.

13.17 The case of Eric Moger

In another case, doctors found a tumor on the left side of the face of Eric Moger, aged 60. Almost half of his face was urgently removed, including the eyes, chin, and cheekbones.

he underwent multiple reconstruction operations, but none were successful. Years later, prosthetic surgeon Andrew Dawood used a 3-D printer to make a prosthesis for facial reconstruction and filled the exposed area on the face of Eric. This procedure achieved good results.

13.18 The case of Kaiba Gionfriddo

Another well-known case is that of Kaiba Gionfriddo. When he was born, his trachea cannot be supported, causing breathing difficulties. Then, a medical team used 3-D printing to make a prosthesis, simulating the path of his trachea and bronchi. The operation was very successful, and only 21 days after the operation, he no longer required the help of respiratory equipment.

Health technology is becoming increasingly important. The benefits are multifaceted and cannot be ignored: improving hospital management, speeding up diagnosis, and improving the quality of life of patients. With 3-D printing technology, the chances of error can be reduced, increasing the possibilities of an ideal treatment option for all patients. In conclusion, there is no doubt that the use of 3-D printing is a promising trend for the future.

References

Aquino, R. P., Barile, S., Grasso, A., & Saviano, M. (2018). Envisioning smart and sustainable healthcare: 3D printing technologies for personalized medication. *Futures, 103*, 35–50.

Eshkalak, S. K., Ghomi, E. R., Dai, Y., Choudhury, D., & Ramakrishna, S. (2020). The role of three-dimensional printing in healthcare and medicine. *Materials & Design*, 108940.

Hornick, J. (2017). 3D printing in healthcare. *Journal of 3D Printing in Medicine, 1*(1), 13–17.

Trenfield, S. J., Awad, A., Goyanes, A., Gaisford, S., & Basit, A. W. (2018). 3D printing pharmaceuticals: Drug development to frontline care. *Trends in Pharmacological Sciences, 39*(5), 440–451.

Trenfield, S. J., Awad, A., Madla, C. M., Hatton, G. B., Firth, J., Goyanes, A., ... Basit, A. W. (2019). Shaping the future: Recent advances of 3D printing in drug delivery and healthcare. *Expert Opinion on Drug Delivery, 16*(10), 1081–1094.

Further reading

4 Ways 3D Printing is Transforming the Maritime Industry. (2017). Retrieved 26 February 2021 from: https://www.linkedin.com/pulse/4-ways-3d-printing-transforming-maritime-industry-chuck-fox.

CHAPTER 14

3-D printing in the hearing aid industry

Contents

14.1 Introduction

An important fact of life is that, as we age, our physical abilities/functions depreciate to a certain extent, including, hearing and vision. However, it is not only age that causes this depreciation, but also other factors. Packer (2016) pointed out that the history of hearing loss can be traced back hundreds of years, and efforts to correct it have also existed for hundreds of years ago. However, further research and investigation on hearing aids are needed to help overcome the remaining obstacles to hearing restoration.

It has been recognized that hearing impairment can occur in completely different ways, namely conductive hearing, sensorineural hearing loss, and mixed hearing loss. Sensorineural hearing loss mainly occurs when the nerves in the ear are damaged. This is mainly due to aging and noise damage to the passages inside the ears. However, this can be treated and resolved with hearing aids. The second type is called conductive hearing loss. This particular condition is caused by obstacles in the outer ear, including earwax, tumors, and body fluids. This kind of hearing loss can be treated with surgery. Finally, there is mixed hearing loss: this is caused by

3D Printing
ISBN 978-0-323-99463-7
https://doi.org/10.1016/B978-0-323-99463-7.00014-1

conductive and sensorineural hearing loss. Starkey (2021) carried out a review to determine the extent of hearing loss that needs to be certified by medical personnel.

In addition to different types of hearing loss, it also occurs to varying degrees. Reflecting the degree of hearing loss clinically, four levels have been determined. For example, there is moderate hearing loss, in which, when people speak in a normal tone, people find it difficult to hear the speech. The second is mild hearing loss, in which it is difficult for a person to hear soft sounds, but they can hear some medium-pitched sounds. The third degree is severe hearing loss, when the patient hears only loud but not normal sounds. Finally, there is severe hearing loss, in which case the individual will hear very loud sounds but none that are medium or normal volumes (Starkey, 2016).

14.2 What

After gaining a clear understanding of hearing problems and their symptoms, it was realized that the problem is growing, and so there have been many innovations and attempts to combat hearing loss since ancient times. As the struggle to remedy hearing loss and deafness has continued, recent records indicate that approximately 13% of American citizens have hearing loss in both ears (Butler, 2015). The most important point to note is that the technology to improve this problem is underway, and understanding its history is the key to determining the development of modern hearing aids.

The first stage of this influence is called the preindustrial revolution. It is claimed to have started before the 19th century, and gave birth to so-called acoustic amplification. By the 19th century, technological developments remained very slow, and acoustics were used for this effect which relied on mechanical equipment to amplify the sound of speech (Butler, 2015). At that time, ear speakers were the most popular items used to assist hearing. These collect voice waves and redirect them to the ears. To make matters worse, this kind of hearing aid only amplifies the sound but cannot solve the real underlying problem.

14.2.1 Before the industrial revolution

The first electronic hearing aid was invented at the second stage of their evolution. These were inspired by the invention of the telephone. Many scientists in the industry are working hard to replicate the functions of mobile phones and assimilate them into hearing aids. For example, Thomas

Edison tried to use a special transmitter to enhance electronic telephone signals, but the results were not optimistic (Butler, 2015). Soon after this invention, a vacuum tube technology was invented, which provided up to 70 dB greater amplification in the presence of limited auditory feedback destruction (Butler, 2015). The user needs to direct the vacuum toward the receiver. Nevertheless, since the vacuum tube was heavy, weighing more than 200 pounds, it was inconvenient to carry and its usefulness thus limited, giving way to other inventions.

After the limited portability of the vacuum, the birth of wearable hearing aids came into being. It is worth noting that the first wearable hearing aid worn by the user was released in 1998 (Butler, 2015). In the 1950s, a gateway technology called transistors was used as a more convenient hearing aid for use with hearing aids, which later became common. However, the first ear-only hearing aid was discovered later, and was much better than the previous models, being smaller and more portable. However, it should be noted that although there have been various improvements to hearing aids, these devices still have problems with poor sound quality and auditory feedback.

This situation led to the latest stage in the development of hearing aids. This trend is labeled as "the most advanced hearing aid technology" (Butler, 2015). At this stage of development, customer choice dominates. This technology also reduces the incidence of poor sound quality and auditory feedback problems that were witnessed in the earlier stages. Today, digital hearing aids account for the largest market share; they have become smart and sensitive enough to produce good sound quality and eliminate background noise. They also have features such as directional listening, which can capture only the sound the user wishes to hear.

In contrast to the above technologies, the current 3-D-printed hearing aids have significant advantages in replicating the natural sounds of the environment. In addition, the current technological advancements in this area include the 3-D printing of hearing aids. This chapter discusses the six w's, what, when, where, why, who, and how of 3-D printing popularity in the hearing aid industry.

This is a relatively new adaptive technology that involves the construction of dimensional objects from different 3-D models in the digital world. 3-D printing refers to the process of solidifying and connecting different objects to produce unusual components. In the 1980s, it was used more often for rapid prototyping. By 2019, 3-D printing had been applied in various industrial production technologies. This highly developed manufacturing technology has gained increased applications in the aerospace, defense industry, and most importantly health care, including remodeling operations (Dodziuk, 2016). An eyewitness noted that despite attempts in the late 1980s and early 1990s, a company that stuck to traditional methods survived (Ruta et al., 2019).

14.3 Where

Hearing loss occurs when the ear drum stops extracting surrounding sounds, or the inner ear is blocked or damaged, which can cause partial deafness. As for the aforementioned types of hearing loss, their causes are consistent with the research of Starkey (2021). Sensorineural hearing loss can be caused by aging, shingles, stroke, obesity, high blood pressure, diabetes, injury, meningitis, high fever or high body temperature, acoustic tumors, ototoxic drugs, smoking, excessive exposure to noise, and genetic and viral infections, for example, measles or mumps. On the contrary, conductive hearing loss is caused by tumors or abnormal growth, earwax accumulation, eardrum scars or sharp foreign bodies in the ear canal, abnormal middle ear hearing, middle ear bone dislocation, and finally tympanic membrane scars or perforations. Hearing loss has clear symptoms, such as when the patient requests speech to be repeated, even if the communication was audible. However, medical analysis has proposed and tried to show how these symptoms exist in different dimensions (NHS, 2018). Starting with the monaural effect, these signs will prevail when the sound becomes clearer, or when speech and sound happen to be quieter than normal. One gets nervous when he or she ignores background sounds or differentiate between different sounds, and his or her hearing gets worse when one hears sounds coming from one side. In young children and babies hearing loss can be detected when they recognize people by sight rather than by sound.

14.4 Who

In October 2001, Topholm asked scholars to develop some software, which later gave birth to 3Shape. In 2002, Widex began to apply 3-D printing in Iceland and some other countries. A clinic was opened in Denmark in February 2003, and the United States followed up in February 2003. Recently, the application of 3-D technology has begun to be applied on the market. According to Phil Reeves, author of the 3-D Printing

Industry Report, there are more than 10,000,000 3-D-printed hearing aids in circulation worldwide (Sharma, 2013). In addition, Jenna, the marketing assistant of Envision Technology, believes that most hearing aids in the world are manufactured using 3-D printers, and the reason behind this innovation is that 3-D printers have turned a physical, labor-intensive industry into an automated industry.

14.5 Why

The adoption of this technology has highlighted its advantages and advantages over other technologies. There are many advantages to using 3-D-printed hearing aids. One of the benefits is the customization factor. This is very necessary as no two ear canals or ears are exactly the same. Therefore it was almost impossible to use the old manufacturing process to create hearing aids that were suitable for everyone. This explains the impact of 3-D technology on the hearing-impaired and medical society (Todays Hearing TM, 2021). Therefore, what was once considered a labor-intensive process is now a widely automated and efficient process because it links technology with practicality and perfection.

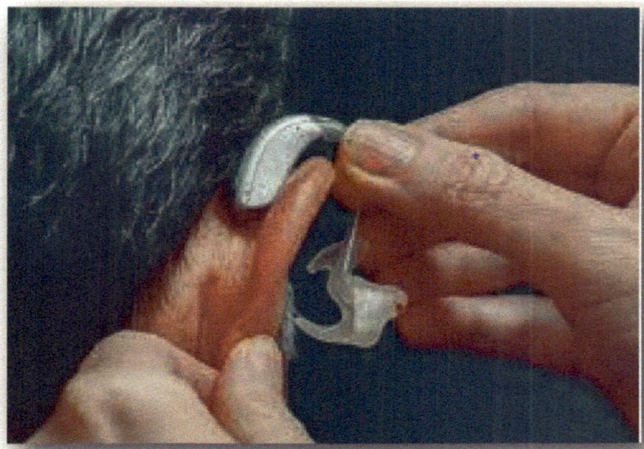

Illustration of a 3-D hearing aid tool. *(From Getty images. (2020).)*

For example, at the start of this era, Sonova became the first company in the world to use 3-D printing technology for production, and recently, this technology has become the benchmark of Sonova. All shell-shaped custom earpieces and custom in-ear hearing aids are products of 3D technology, which enable the hearing aids to optimally fit the customer's ear canal and meet the needs of the customer's degree of hearing loss (Sonova Hearing World, 2021).

Compared with traditional bulk devices with batteries, the cost of hearing aid devices has been reduced. Scientists believe that 3-D-printed hearing aids that cost no more than a dollar will make hearing aids available to those who cannot currently afford them (New Scientist, 2019). An NGO called World Wide Hearing has stated that in developing countries, fewer than 40% of people in need of hearing aids can afford one, also, there are cheaper amplification devices that can simply amplify all sounds to help with mild hearing loss. Therefore, through this evidence, we can conclude that 3-D printing technology simplifies the access to and quality of hearing aid devices.

Compared with bulky traditional hearing aids, this technology can produce more comfortable hearing aids. This method can help people all over the world to hear more comfortably. As 35 million American citizens face the problem of hearing impairment, a more comfortable method of wearing of wearing hearing aids is required, and this can only be achieved with superior hearing aids that meet their needs, as shown in the figure below.

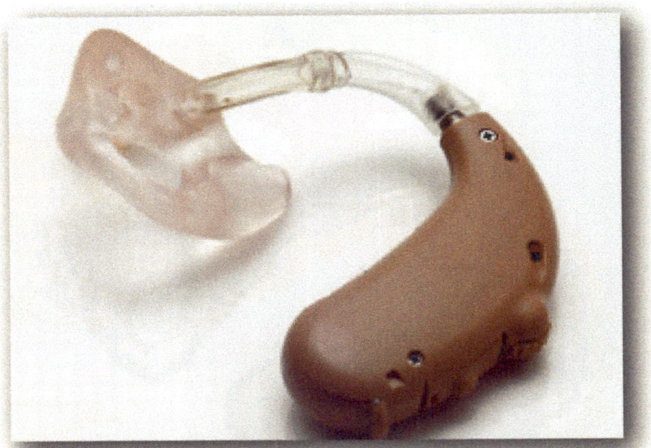

(Getty images. (2020). Hearing aid premium high res photos. *https://www.gettyimages.com/photos/hearing-aid?phrase=hearing%20aid&sort=mostpopular.)*

Another important factor is production speed when 3-D printing is used to produce hearing aids. For example, the 10 million 3-D-printed hearing aids currently on the US market only required 24 hours for production. This extraordinary process has completely changed the way that products are made. In the past, hearing aids took much longer to produce, and they were still not fully customized. However, we are fortunate that those long and difficult manufacturing processes no longer exist thanks to the new era brought about by 3-D printing technology.

Finally, comfort and performance are the main reasons why 3-D printing is so important. Printed hearing aids provide customers with a greater degree of comfort, and are tailor-made for audiologists to suit the individual problem and allow reasonable adjustments. For designs that are worn in the ear all day long, getting a perfect fit is especially important. In this development, comfort resonates with performance, because customized hearing aids provide a perfect fit. They also resonate with advanced auditory experiences. Most importantly, the sound does not leak from the receiver in the ear back to the original microphone. In addition to comfort, another important requirement for hearing aids is durability. Given the customization of these products, a perfect fit will only depends on the result in the accuracies of the production of equipment.

Although 3-D hearing aids have many benefits, they still face some challenges for most companies. The biggest anticipated problem is related to the re-education of the labor force, as using software and 3-D printers requires a number of skills, and all hearing aid companies claim that this is the biggest challenge they face. Although the process requires work, many companies say that this change is not as drastic as they had expected. In many companies, such as Widex, Phonak, and Oticon, technicians are relatively confident in this change and are eager to study this new technology. The removal of toxic fumes appears to be an important factor in this uncertainty. In addition, a large network of artisan skills remains, because the manufacture of hearing aids requires both manual and visual skills. Therefore, compared with nonartisans, it is simpler for well-trained craftsmen to learn and adapt to the use of 3-D printers.

In terms of durability, some chemicals used for personal hygiene can damage the casing of the device, creating holes that can affect the functionality of the device. For example, these devices may shorten their useful life. Therefore, hearing aid manufacturers dealing with these issues should focus on obtaining durable materials for maximum and long-lasting performance, that is, materials that meet durability, comfort, and chemical resistance requirements as a whole.

14.6 When

It has been a long time since a team of engineers first conducted research in the field of hearing aid development to see if it was possible to use 3-D printing technology to manufacture hearing shells. The main goal of the team was to get rid of small-scale production and maximize industrialization and control. The study concluded that, while possible, the cost was too high, with an estimated capital investment of around $4-5 million. This leaves applications and adoption in the air for years to come. It was not until 1992 that the CEO of the Danish family-owned company Widex was granted a patent for the use of 3D printing technology, making it a name in the hearing aid industry (Sandström, 2015).

As there are many patents for stereolithography technology, there is no unique patent on the manufacture of hearing aid housings. In the following year, Widex began to produce hearing aid housings, and received support from many competitors. Siemens and Phonak made a further adaptation in the late 1990s. The company adopted more 3-D dynamics (Sandström, 2015), and in 2000, Siemens bosses were convinced that this attempt would be very useful. It started to promote 3-D manufacturing, focusing on efficiency and speed. After this great innovation, Siemens decided to work with Phonak to reduce costs and circumvent the more expensive components in the 3-D printing invention. In 1999, when they started to explore 3-D printing technology, they began to provide good biocompatible materials, resulting in the ability to produce functional hearing aid shells.

14.7 How

3-D printing has received the attention it deserves; the production of hearing aids has been helped by the 3-D printing process for some time. This is an important method because these devices have highly personalized characteristics to suit each individual's ears.

These hearing aids were created using an adaptive mechanism suitable for everyone to ensure high user comfort. Using 3-D laser scanning, the whole process can be completed in less than 1 day. There are differences in the technology used by different companies, for example, an audiologist might first obtain an image of the ear using a laser scanner and then obtain a client's file from the internet cloud.

After completing the scan and quality check, the hearing aid model can be produced. Through the printer, the user can obtain the accurate shape of the hearing aid from the ventilation holes. Appropriate acoustics and other components are applied also to obtain the proper fit (Todays Hearing TM, 2021). After determining the quality and efficiency, the hearing aid stick is placed in the housing to make the sound. Therefore, this technology has undoubtedly improved hearing aid technology, which explains why more than 10 million people use 3-D printed hearing aids currently.

14.8 Conclusion

There is no doubt that 3-D technology has played an important role in improving the hearing aid industry. It makes the production of customized ear shells perfect. In addition to the portability of the device for the user, the product is widely available and accessible. World Wide Hearing stated that less than 40% of people in developing countries who need hearing aids can afford them, also, there are cheaper sound amplification devices that can simply amplify all sounds to help with mild hearing loss. Therefore, through this evidence, it can be concluded that 3-D printing technology simplifies the use and quality of hearing aids.

References

Butler, I. (2015). *The evolution of hearing aid technology*. https://myhearingcenters.com/blog/the-evolution-of-hearing-aid-technology/#: ~ :text=The%20first%20wearable%20hearing%20aid,of%20a%20pair%20of%20glasses.

Dodziuk, H. (2016). *Applications of 3D printing in healthcare*. https://europepmc.org/article/med/27785150.

Getty images. (2020). *Hearing aid premium high res photos*. https://www.gettyimages.com/photos/hearing-aid?phrase=hearing%20aid&sort=mostpopular.

New Scientist. (2019). *Hearing aid uses 3D printed parts and costs less than a dollar*. https://www.newscientist.com/article/2222107-hearing-aid-uses-3d-printed-parts-and-costs-less-than-a-dollar/.

NHS. (August 1, 2018). *Symptoms-hearing loss*. https://www.nhs.uk/conditions/hearing-loss/symptoms/.

Packer, L. (2016). *Hearing aid history: From ear trumpets to digital technology*. https://www.healthyhearing.com/report/47717-Digital-hearing-aid-history.

Ruta et al. (2019). https://www.USitc.gov (March, 2021). US International Trade Commission.

Sandström, C. (2015). *Adopting 3D Printing for manufacturing — the case of the hearing aid industry*. https://ratio.se/app/uploads/2015/12/cs_3dprinting_hearingaid_262.pdf.

Sharma, R. (2013). *The 3D printing revolution you have not heard about*. Forboes. https://www.forbes.com/sites/rakeshsharma/2013/07/08/the-3d-printing-revolution-you-have-not-heard-about/?sh=7b6ab1801a6b.

Sonova Hear the World. (2021). *3D printing technology for improved hearing*. https://www.sonova.com/en/story/innovation/3d-printing-technology-improved-hearing.
Starkey. (March 1, 2016). https://www.Starkey.com. Minneapolis, USA.
Starkey. (2021). *Types and causes of hearing loss*. https://www.starkey.com/hearing-loss/types-and-causes.
Todays Hearing TM. (2021). *How 3D printers are revolutionizing the hearing aid industry*. https://tohear.com/how-3d-printers-are-revolutionizing-the-hearing-aid-industry/.

Further reading

Freund, C., Mulabdic, A., & Ruta, M. (2019). *Is 3D printing a threat to global trade? The trade effects you didn't hear about*. http://documents1.worldbank.org/curated/en/152701569432061451/pdf/Is-3D-Printing-a-Threat-to-Global-Trade-The-Trade-Effects-You-Didnt-Hear-About.pdf.

CHAPTER 15

3-D printing in the maritime industry

Contents

15.1 What

The shipbuilding industry includes engineering that deals with ship development, as well as the industrial branch that completes and repairs ships. In all areas, 3-D printers can simplify tasks, shorten processes, and save costs (Kostidi & Nikitakos, 2017). Manufacturing using 3-D printing is increasingly entering shipyards and related factories. Increasing numbers of shipbuilding companies are discovering for themselves the potential and possibilities of 3-D printing technology, such as less manufacturing wear or faster prototyping. 3-D printing, also known as additive manufacturing, may have a significant impact on profitable after-sales business, product development, and the competitive landscape of shipbuilders (Tas & Sener, 2019). At the same time, it has opened up various opportunities for strengthening the existing market position and tapping into the potential for new markets.

3D Printing
ISBN 978-0-323-99463-7
https://doi.org/10.1016/B978-0-323-99463-7.00015-3

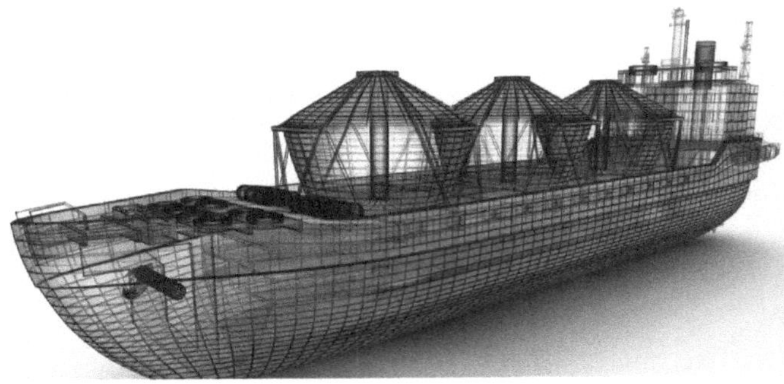

(From: 4 Ways 3-D Printing is Transforming the Maritime Industry. (2017). Retrieved February 26, 2021, from https://www.linkedin.com/pulse/4-ways-3D-printing-transforming-maritime-industry-chuck-fox.)

Therefore, the managers of such manufacturers should respond to the opportunities and changes in the upcoming market environment and take necessary measures at the same time. There is no doubt that the existing business model and market conditions are changing with 3-D printing in traditional production lines. This will affect particularly technically complex products and small batch production, enabling 3-D printing to be effectively used in the shipbuilding industry.

15.2 Prototypes, interiors, spare parts, tools

A full-size ship has many complicated accessories, which can be complicated geometric shapes. Many of these parts must be manufactured to scale, and the traditional method of making prototypes from plastic or wood is a very labor-intensive and demanding process. This may take several weeks, and is also quite expensive. With the use of 3-D printers, this situation has changed, with such prototypes and even functional spare parts being produced within 24 hours (Kostidi & Nikitakos, 2018). This allows shipyards to manufacture customized prototype parts for ships at low cost and with significant time savings.

15.3 Large ship parts from industrial 3-D printers

Compared with the previous traditional production methods, the shorter production time and lower installation workload can meet customer requirements more flexibly. 3-D printing in the shipbuilding industry also facilitates the production of propeller pod shells, which are often difficult to produce and assemble. The possibility of printing different parts using only a 3-D printer makes assembly easier.

15.4 When

In recent years, many industries have fully realized the benefits of additive technology and have effectively integrated them into the manufacturing process (Vujović et al., 2018). More and more companies in the aerospace, automotive, engineering, oil and gas, medical, and jewelry industries around the world are using incredible opportunities presented by 3-D printing to open up promising new methods.

Shipbuilding is a conservative industry, but it is inevitable that it will enter the 3-D world. Three-dimensional scanning and printing have unique potential because they can reproduce the most complex spatial forms and objects. So far, the application of 3-D printing technology for ship parts is considered a good idea, but its practicality and cost-effectiveness have not been widely recognized.

(From: RAMLAB Opens 3-D Printing Laboratory for Maritime Industry, All3-DP. (2021). Retrieved February 26, 2021, from https://all3Dp.com/ramlab-opens-worlds-first-3D-printing-laboratory-for-maritime-industry/.)

15.5 Where

As in the past, and even more so now. Hypotheses derived from novelty and fascination are being made due to the exponential progress in the state of the art technogies. These hypotheses indicate that the full-scale transformation of the world we live in has unlimited possibilities. The road of

3-D printing progress has been is arduous; printing technology began its development 30 years ago, and the impact has not reached what was initially imagined, as it require a high level of computing power and ability to generate 3-D plans for printers to use.

Other current disadvantages are that, in many cases, the surface finish of printed products is very rough, requiring additional polishing work, and even the operation of the printer is not fully automatic, requiring specialization and a certain amount of manual work. Although new methods and materials are being developed, it is impossible to use different materials for printing at the same time.

It is also clear that the progress of using 3-D printing technology is parallel to the development of new materials (not just polymers) that can withstand the strong physical stresses that must be withstood.

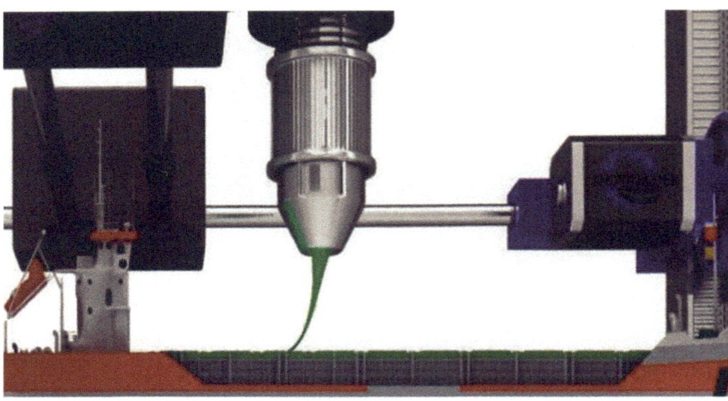

(From: Hanaphy, P., Hanaphy, P., Everett, H., Everett, H., Sertoglu, K., & Sertoglu, K. et al. (2016). 3D printing rocks the boat in the Netherlands as NISS shipbuilders provide perspective - 3D Printing Industry. *Retrieved February 26, 2021, from https://3dprintingindustry.com/news/3d-printing-rocks-boat-netherlands-niss-ship-builders-provide-perspective-97873/.)*

Some analysts believe that, at least for the next 2 decades, the impact of 3-D printing technology on maritime transportation will not be significant. They believe that the overall increase in transportation demand brought about by continued growth will offset these effects. Others believe that the impact of these new digital technologies will be revolutionary, and that it will drastically change industry scenarios and transportation modes by bringing production closer to the center of demand rather than dispersing to places where labor costs are low, thereby significantly affecting shipping and transportation modes.

Some have even discussed about the future "factory ship" with huge potential for 3-D printers, which could partially replace the proliferation of these machines in consumer centers. The greater the result of the changes that will occur, the faster the technological developments required will advance.

However, in general, technical problems seem to be easier to solve than those caused by the business environment: intellectual property rights, licensing, and the fight against piracy in this case, etc. seem to be more difficult to solve, because these have always been unstoppable since the "digital world" In any case, it remains to be seen whether there will be more opportunities or challenges for the global transportation concept.

15.6 Who

We have long been accustomed to modern technology replacing the ubiquitous old methods requiring physical and mental labor. The freight field is no exception. Packages are now delivered by robots, who control the ships; but who ever imagined that the robot itself would become outdated! Before our eyes, new technologies are changing the production process itself and the supply chain that has developed over the years, revealing new forms and decommissioning old forms. According to scientists, the latter threatens the foundation of international trade shipping and trouble may come from (or rather will come from) places they hadn't anticipated. According to a joint study published by scientists from Nanyang Technological University (NTU) and National University of Singapore (NUS) at the Maritime Research Institute of Singapore meeting, 3-D printing will become the killer of the modern trucking industry.

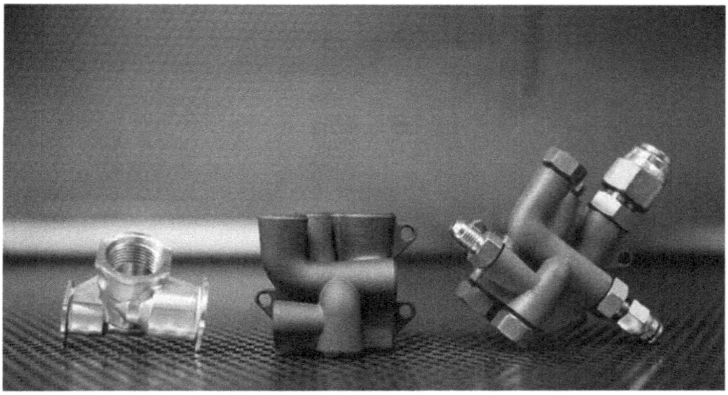

(From: (2017). TOP 10 3D printing applications in maritime - 3dnatives. *Retrieved February 26, 2021, from https://www.3dnatives.com/en/3d-printing-maritime-applications230820174/.)*

According to these researchers, these changes are not far from happening. And in fact, they are already happening in some circumstances, and 3-D printing will have an impact on the global logistics market in the next 20 years. It is estimated that by 2040–2045, the popularization of additive technology could reduce shipping demand by 10–20%! At the same time, according to scientists, the increase in the transportation demand for 3-D printing raw materials will not compensate for the decrease in the delivery of finished products. Additionally, significantly reducing additive manufacturing losses will significantly reduce the need for bulk shipping.

15.7 Why

3-D printing of spare parts can bring value to the shipping industry. The shipping industry is generally conservative in terms of change, but 3-D printing could be used well in this industry. Since ships are often far from maintenance facilities and spare parts warehouses, if spare parts can be produced directly on ships, the distance problem can be easily solved.

The shipping industry is no stranger to 3-D printing, and some successful application cases have already been mentioned. For example, in 2016, the "3-D printing marine spare parts" project was initiated by the Innovation Zone, the Port of Rotterdam Authority, and RDM Makerspace. A total of 28 companies and institutions participated in the maritime 3-D printing application project. The project results showed that 3-D printing is expected to print a variety of maritime spare parts, however some adjustments are needed in terms of quality. The author also mentioned another project involving the world's first approved 3-D-printed hull propeller.

The U.S. Navy has used 3-D printing in many maintenance tasks with success. Researchers claim that maintenance work will allow the U.S. Navy to install and test 3-D printers on ships for long periods of time (Garrido et al., 2020). At the same time, the crew will also be busy printing plastic syringes, fuel tank caps, and model aircraft that simulate the cockpit. The U.S. Navy claims that it will take several years to print out actual ship spare parts, but this is a very good start.

Those working in the maritime industry are mainly concerned about the quality of additive manufacturing spare parts. Another problem is the cost of the 3-D printer and the cost of parts processing.

Although the problem has not been resolved, exploring 3-D printing at sea continues, with the best solution for 3-D printing on board, protecting intellectual property rights, and training technicians being addressed. However, in general, 3-D printing is a "promising" technology that the maritime industry should seriously consider. So far, the application of 3-D printing technology for ship parts has been considered a good idea. Nevertheless, its practicality and cost-effectiveness have not yet been widely recognized.

Additive manufacturing technology, commonly referred to as 3-D printing, which provides huge potential value for the maritime industry, especially when it comes to the supply of replacement parts. It is fast and not restricted by geography, thus reducing the number of parts; fast replacement impact on the safety and efficiency of the ship's sailing. More importantly, like many others, more and more people are turning 3D printing into practice, not just on paper.

15.8 How

The current situation makes illustrates the impact that the new 3-D printing technology may have on the industrial and marine worlds, affecting the shipbuilding and repair industry, as well as the shipping industry, shipping and maritime traffic, and even the military industry.

An example is the new General Electric Aviation factory which uses a 3-D printer to manufacture fuel nozzles for the reactor of the Airbus 320 aircraft, from Siemens gas turbines. There are other problems related to the possibility of applying these technologies to solve various types of medical operations.

Essentially, the technology used makes it possible to convert newly designed virtual 3-D computer files and real objects digitized by a 3-D scanner into tangible physical objects.

First, the use of this technology could pose a huge advantage: significant reduction in the chain and logistics space, not only involving the storage of manufactured items (spare parts, manufacturing intermediate components or other types); including the transfer from the place of production to that of use or consumption; location of transportation (if applicable) and the time spent in these operations. Another advantage is the elimination of waste or manufacturing surplus. This more important with higher value materials.

Obviously, today, the more advanced and complex the products manufactured, the more sensitive and complex the supply chain structure, involving the collaboration and coordination of many different participants as well as suppliers and distributors. A schematic explanation of this, especially in the case of complex products, can be expressed as follows.

Traditional chain: raw material > intermediate elements > equipment > final manufacturing > distribution > seller > buyer/end consumer.

Under the assumption of making full use of 3-D printing technology, this chain could look like this: raw material > final manufacturing and final buyer/consumer.

In a traditional chain, all the spaces indicated by arrows, in addition to other aspects, also require transportation, storage, and complex coordination, and usually require partial design. In the second case, the only items transported are raw materials, and the final manufacturer will have a 3-D printer and computer files containing the information needed to digitize the design.

In terms of simplification, computer files will be transmitted over a network, rather than from different geographical points. The traditional costs of transportation, inventory financing, handling, and storage will actually be eliminated. In the case of discontinuous or small batch manufacturing, the advantages of 3-D technology will be greater, because in most cases, depending on the capacity and flexibility of the printing press, the only thing that needs to be modified is the files for computers.

It seems logical in this type of evolution that the design will also be modified so that the information digitized to the printing press is better adapted to the working methods and guidelines. Therefore, we can infer that we are facing an industrial revolution, the consequences of which are currently unpredictable, but it may completely change the world's production and consumption habits and affect more traditional-based field manufacturing types, especially when wage costs are very Low populations need storage and transport logistics chains to export and sell their products.

This future will change the behavior of society, industry, and individuals. It is not yet possible to predict to what extent, but it seems logical to consider some key factors. The applied technology will develop faster, which is the normal state of the "information society" These solutions will be very different from the solutions applicable today. Digitization allows us to engage in more global and interconnected activities. Logistics will change. Considering that the entire system can be copied anywhere, we can make objects wherever we have a suitable 3-D printer and the required software. All of these raise important and multiple questions about the concepts of intellectual property and industrial property rights; even today, while we are still advancing at a more basic level, these unknown problems have not been resolved. Problems caused by dominance, especially those related to software ownership, can have huge economic and social impacts.

With all that we know today (which is very little), it can be expected that the technology we are talking about will affect the industry and transportation services and global logistics networks, and in our case, will affect shipping. The sea transportation of raw materials and food will not be affected. They will continue to be necessary, and will only change as the global economy grows, as has always happened. On the other hand, some modes of maritime transportation of raw materials may change, because people who receive finished products now, in the future will need raw materials to make their previous products "in situ"

The transportation of energy products will not change due to the development of this technology. It will only do so in terms of energy changes, and it is very important because renewable energy is replacing fossil fuels, although it is expected to be very slow on a global scale. (It is one thing to stop the growth of fossil fuel use, but it is another to drastically reduce its use in the short term).

On the contrary, the impact on container transportation is a basic element of finished goods logistics, which may affect containers, multipurpose, and general cargo ships.

References

Garrido, J., Sáez, J., Armesto, J. I., Espada, A. M., Silva, D., Goikoetxea, J., Arillaga, B., & Lekube, B. (2020). 3D printing as an enabling technology to implement maritime plastic circular economy. *Procedia Manufacturing, 51*, 635–641.

Kostidi, E., & Nikitakos, N. (2017). Exploring the potential of 3D printing of the spare parts supply chain in the maritime industry. In *Safety of sea transportation* (pp. 171–178). London, UK: Informa UK Limited.

Kostidi, E., & Nikitakos, N. (2018). Is it time for the maritime industry to embrace 3d printed spare parts? *TransNav: International Journal on Marine Navigation and Safety of Sea Transportation, 12*.

Tas, S. O., & Sener, B. (2019). The use of additive manufacturing in maritime industry. *International Journal of Engineering Trends and Technology, 67*(6).

Vujović, I., Kuzmanić, I., Kulenović, Z., Maleš, N., & Navy, C. (January 2018). Additive manufacturing of spare parts in maritime industry-technology transfer in mariner education. In *21st international research/expert conference "trends in the development of machinery and associated technology"*.

CHAPTER 16

3-D printing in the mechanics industry

Contents

16.1 What

Since its inception, 3-D printing has continued to grow and develop. It is usually given the somewhat unusual name of additive manufacturing; some even claim that this is a revolutionary process. The 3-D revolution is everywhere in the media currently, and the general public is very aware of its presence. For the general public, this is an interesting topic, but for manufacturers, it is also a revolution, sometimes described as the "new industrial age" In this revolution, it is no longer the manufacturers who make the products, but the consumers themselves.

New opportunities have emerged after the development and implementation of 3-D equipment in the mechanical engineering industry. It is not ubiquitous, but in several areas, additive technology has begun to rapidly replace traditional production methods. In mechanical engineering, 3-D printing creates conditions that allow for efficient, fast, and effective solutions to various tasks. Among these tasks are prototype development and the production of new parts and assemblies (concept models, test samples); creating more modern systems and their elements; and repairing and replacing old parts. In practice, 3-D equipment has shown its effectiveness and ability to produce better quality and lower cost products (Ramya & Vanapalli, 2016).

3D Printing
ISBN 978-0-323-99463-7
https://doi.org/10.1016/B978-0-323-99463-7.00016-5

16.2 Why

Mechanical engineers can obtain almost any object by entrusting the project to a 3-D printing company like TH Industries. According to the specific situation of the project, the experts from a 3-D printing company will choose the computer-aided design (CAD) software or construction software suitable for creating 3-D files. Using 3-D modeling software with excellent analysis and visualization options can help to create a complete picture of any project. This is especially useful when dealing with fairly complex mechanical parts. With 3-D printing, the topology of the parts to be designed can be optimized to improve the project. Specifically, this process can find the best compromise between resistance, weight, and production cost for mechanical parts. By using the 3-D printing service provided by TH Industries for a project, the most material suitable for the project can be chosen from the variety of 3-D printing materials they provide (Ramya & Vanapalli, 2016).

Thanks to 3-D files and selected materials (PLA plastic, metal, ABS plastic, etc.), 3-D printers speed up the manufacturing process of mechanical parts. This technology allows rapid prototyping and can perform the required testing in less than 2 seconds. 3-D printing allows the most familiar mechanical parts that traditionally consist of many components to be mass produced. This reduces the assembly and welding steps. This manufacturing process will save time, allowing faster production. Today, additive manufacturing enables mechanical engineers to design, test, build, and operate mechanical, hydraulic, and thermal machines (Ramya & Vanapalli, 2016).

In addition to automotive engineering, military, maritime, and aerospace engineering are all sub-branches of mechanical engineering that benefit from the advantages of 3-D printing. For example, in the aviation field, GE aircraft fuel nozzles designed by engineers consist of no less than 25 different components. Today, 3-D printing allows these nozzles to be made in one piece. This increases the strength of the device fivefold. In addition, the time normally spent installing nozzles is also eliminated. The quality control process is completed very quickly, and would normally take several months. What a time saver! Therefore, 3-D printing has greatly affected the field of mechanical engineering because it has completely changed almost all processes traditionally used in this field. 3-D printers can perform rapid prototyping and save manufacturers a great deal of time and money. For a market that faces fierce competition such as automobiles, the time saved in prototyping parts (engines, spare parts) represents a real competitive advantage.

Another big advantage of 3-D printers is that they can customize parts. One of the most common applications used by manufacturers is rapid prototyping. In all-round development, 3-D printing has developed rapidly. Large parts can now be printed in a variety of materials, including plastic, wood, resin, textiles, and some metals (Chong et al., 2018).

16.3 Where

Three-dimensional computer modeling has been firmly established in many fields of production, including construction, energy, film production, and scientific activities. Computer modeling and natural reproduction of objects in 3-D form have been actively introduced into various fields of human life. This method of manufacturing objects significantly saves materials, energy, and time resources. The rapid development of additive technology methods has reduced the cost of 3-D printers and consumables.

The selection, terminology, and scope of these materials are constantly expanding. Additive technologies are becoming increasingly common because they allow us to obtain products at a significantly reduced cost compared to manufacturing similar products through traditional technical methods. These technologies make it possible to make major changes and additions to the product design, so as to meet the needs of a wide range of customers. These technologies can achieve a very short technology cycle, provide automated control of manufacturing operations, and require the least production time. General Electric Company of the United States is a pioneer in applying this technology to the production of industrial products.

16.4 When

In the manufacture of very expensive parts such as engine turbine blades, in the traditional method—through precision investment casting—the development of the process takes 4–5 weeks, and the production of parts using 3-D printing can reduce the production time to 1 day. The economical additive technology used to manufacture such parts that require the most accurate execution and the least number of adjustments is about 10 times more profitable than the old traditional technology. The technology of casting blade models with wax or urea is beginning to lose the ability to create these models using 3-D printing. The company's printer for these purposes, the ProJet 3510SD, can provide high-precision models for all stages of production from development work to completion of samples in a short period of time (Chong et al., 2018).

The Chinese automotive electronics manufacturer, TE Connectivity, provides electronic components for many automotive manufacturers around the world. The company's engineers develop and production workers are engaged in the manufacture of multicomponent systems such as sensor systems, on-board computers, and navigation boxes. These products need continuous improvement and also they need to continuously mass-produce cars. In order to speed up the development and manufacturing of such systems and components, the company's management decided to use additive technology, which will enable designers and engineers to quickly obtain new development results and carry out subsequent manufacturing. For these purposes, the Objet 24 3-D printer was chosen. This printer has significantly increased the number of new product developments, system and product prototypes, and design developments. Additive technology has significantly accelerated the development and release of existing and new products, increased profits and the inflow of new capital, and accelerated the process of products conquering new markets.

This new technology is used in the design and creation of the casting master models. The company's customers can see and evaluate prototypes of future products in a complete physical volume, which greatly increases buyers' confidence in the company and increases the demand for its products. The company calculates the cost of equipment purchased for the introduction of additive technology in 1 year. This company's experts now have the opportunity to quickly respond to market changes and product demand, which makes it possible to increase profits, as it is more flexible and liquid than its competitors.

Nippon Katan Corporation, a Japanese electric company that manufactures and installs transmission lines, has also recently applied additive manufacturing to components and assemblies, such as power supply brackets, insulators, and fasteners (Chong et al., 2018).

The company uses a ProJet 3510 HD 3-D printer, which allows the company's experts to visually show their products to customers, increasing the speed and efficiency of product manufacturing. Product prototypes are used to help to improve the design. During the 3-D printing process, the ability of the new design to withstand weather conditions can be quickly adjusted and analyzed: including for rain, cold, snow, ice, strong winds, and thunderstorms.

Additive technology has begun to penetrate into all areas of life. In 1995, a group of NASA experts showed the first large-scale 3-D printer to the public. They were preparing to use it to build shelters for scientists and even astronauts. Many construction companies began to correct outdated ideas about construction technology, not only conquering the little-known areas on our planet, but also elsewhere in the universe.

The design of the 3-D architectural printer is very simple, because the products produced with it are large in size and the tolerances for the products are also large, with no stringent precision being required. In principle, it is a framed metal structure, including concrete pumps, concrete pumping hoses, and a software package for monitoring the construction process, with a three-dimensional computer model embedded in its memory. This machine can print concrete buildings while leaving room for the installation of metal reinforcement elements.

Chinese experts were the first to develop architectural 3-D printers. Shanghai has used 3-D printers to build entire residential areas. The working area of an architectural 3-D printer can be up to 150 m long and up to 10 m wide. Using this printer, 6-meter-high buildings can be printed in a few days (Tay et al., 2017).

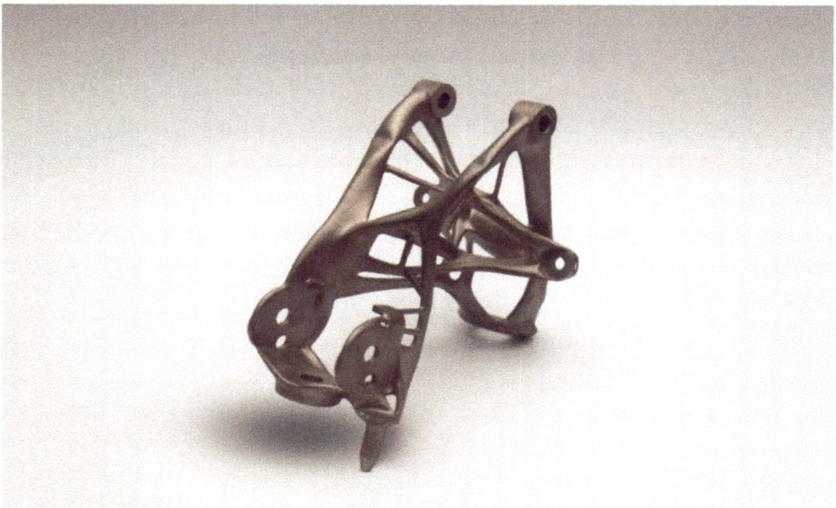

(Source: https://pick3-Dprinter.com/3-D-printer-for-mechanical-parts/.)

16.5 Who

3-D printing technology is most in demand in the production and maintenance of gas turbine generators, as well as the LEAP series aircraft engines that equip Airbus F320neo, Boeing 737 MAX, and COMAC C919 passenger aircraft. General Electric attaches great importance to the production of 3-D printers themselves. One of the latest developments of the company's engineers is the Atlas high-performance inkjet 3-D printer for printing metal products. Another product recently developed by the company's engineers is the Atlas Titan Robotics 3-D printer, which is positioned by the company as the largest 3-D printer using selective laser melting (SLM) technology in the production of parts.

As the number of parts manufactured using 3-D printers increases, the company faces the challenge of accounting and verification of additive manufacturing processes, consumables, and parts. The solution to this problem led to the creation of blockchain technology—a technology used to store and transmit data to create a data archive to verify production. The database contains data related to manufacturing processes involving parts manufactured using additive technology (Wimpenny et al., 2017).

3-D printing allows high- and/or low-quality copies of parts or products to be obtained in the shortest possible time. Using blockchain technology, it is possible to automatically create a file containing the real data passed by a part of the nodes in all operation processes. This information precludes further technical processing, submissions for assembly and sale, violations of technical and production process patterns of parts or assemblies, which means that the control of defective products and measures to segregate defective products in a timely manner can be delegated to various automated equipment, and of course, inseparable from human control.

Using digital technology and special equipment to make products allows the process flow to be reduced while creating products. Nevertheless, at the same time, it does not give a complete picture of the compliance of all the parameters, technical standards, and computer models created by designers of finished products or samples (Tay et al., 2017).

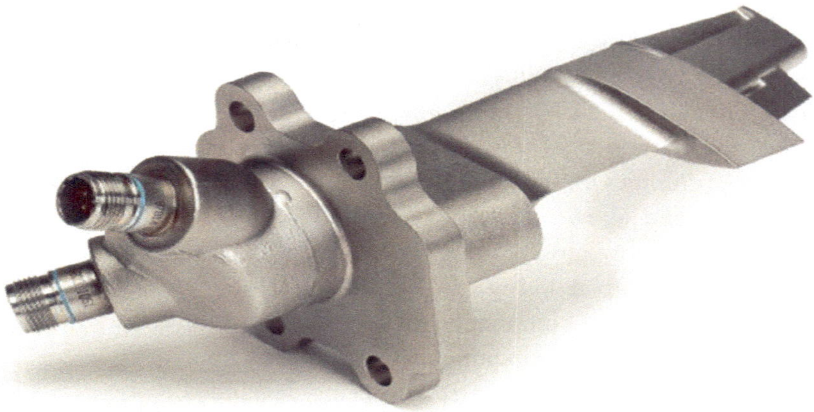

(Source: https://www.sculpteo.com/.)

General Electric experts are studying ways to solve this problem. When using additive technology to convert parts and components for manufacturing, they hope to save $25,000 from each engine produced. Ten percent of the company's products are manufactured using additive technology. Recently, experts from Siemens, NASA, and other industrial giants in the United States, Germany, Japan, and China have actively used additive technology in the production of parts. The American company Turbine Technologies (Wisconsin, USA) and its subsidiary Kutrieb Research specialize in the development and manufacture of turbine engines, actively using additive technology to manufacture the main components of gas turbines, such as blades.

16.6 How

The range of tasks that can be solved with modern 3-D printing systems is expanding almost daily. Products can now be received faster and at a higher quality level. New technologies are expanding the opportunities for engineers and designers in the mechanical engineering industry. With the new equipment, one can more easily solve problems in the process of conceptual design creation and finished product production. 3-D equipment allows the creation of prototypes for testing before mass production starts, and the testing and verification of various features eliminates possible defects in advance (Wimpenny et al., 2017). 3-D printing expands the possibilities of creating instrument housings and equipment components. The new equipment allows the creation of unique objects with unique attributes and ensures the reliable operation of the designed equipment. It also creates ample opportunities for creating convenient and efficient tools to accelerate production, and allows for the creation of casting models quickly and efficiently. It is now possible to make very precise wax, burnt models and silicone casting samples. Using a 3-D printer, one can manufacture products that are ready to run: components of various mechanisms, parts and spare parts for repairs, engine components, etc.

Specifically, the following areas are being developed:

- Selective laser melting (SLM) technology. This technology allows the use of lasers. In contrast, the diameter of the laser beam is very small at about 20 microns, which makes it possible to obtain parts of specified dimensions and surface roughness with sufficiently high accuracy.
- Laser stereolithography (SLA) technology. Under the action of a laser beam, the contours of future products are formed on the surface of a special photosensitive polymer. The product is then immersed in the polymer until a layer thickness is formed. When a projector is used instead of laser (DLP technology), the product is not formed layer by layer, but is immediately formed as a whole. This allows the time to print products to be reduced.
- Selective laser sintering (SLS) technology. In this method, powder materials are used, and products are formed from the powder materials under the action of a laser beam. The molding powder is coated on the substrate, and the product is formed by layering the powder material, while the board is reduced to the thickness of the next layer each time. The material is sintered in the chamber at a temperature slightly lower than the sintering temperature. To prevent oxidation, the indoor atmosphere does not contain oxygen (Mami et al., 2017).
- Electron beam melting (EBM) technology. This technology is used to manufacture products from refractory materials and alloys using electron beams in a vacuum. This method is used to obtain different pure designated chemical compositions and physical mechanical properties (Ramya & Vanapalli, 2016).
- Fused deposition modeling (FDM) technology. This technology involves the use of consumable blank materials (wax, plastic, metal) in the form of threads. The wire is fed into a press by heating and squeezed

onto a cooled work plate in the form of a thin wire by the press. The thread is laid in dense and uniform layers to form a layer of molded product. Then the board is lowered to the thickness of the next layer and the threading operation is repeated (Tay et al., 2017).

- Multiinjection modeling method (MJM) technology. This method replicates the deposition method, replacing the press with a spray head to melt the workpiece into a liquid state.
- Laminating method (LOM) technology. The working material of the blank is in the form of a film, which sticks together when exposed to temperature and pressure. The product is cut from the monolith thus obtained; during this operation, a laser beam or a metal cutting tool is used. Therefore, particles or even pieces of workpiece material may remain on the product, requiring additional cleaning operations.

References

Chong, S., Pan, G. T., Chin, J., Show, P. L., Yang, T. C. K., & Huang, C. M. (2018). Integration of 3D printing and Industry 4.0 into engineering teaching. *Sustainability, 10*(11), 3960.

Mami, F., Revéret, J. P., Fallaha, S., & Margni, M. (2017). Evaluating eco-efficiency of 3D printing in the aeronautic industry. *Journal of Industrial Ecology, 21*(S1), S37–S48.

Ramya, A., & Vanapalli, S. L. (2016). 3D printing technologies in various applications. *International Journal of Mechanical Engineering and Technology,* 7(3), 396–409.

Tay, Y. W. D., Panda, B., Paul, S. C., Noor Mohamed, N. A., Tan, M. J., & Leong, K. F. (2017). 3D printing trends in building and construction industry: A review. *Virtual and Physical Prototyping, 12*(3), 261–276.

Wimpenny, D. I., Pandey, P. M., & Kumar, L. J. (Eds.). (2017). *Advances in 3D printing & additive manufacturing technologies* (pp. 1–195). Singapore: Springer.

CHAPTER 17

3-D printing in the movie industry

Contents

17.1 Introduction

The 3-D printer continues to gain prominence in many areas. It is particularly popular in the cinema industry in order to create the most imaginative sets and accessories. In recent years, Hollywood has seized on this technological innovation to use it in the creation of films. Andrew Clément, a props designer and special effects expert, describes using 3-D printing on a daily basis. This tool has completely changed the way he designs accessories.

17.2 What

The cinema could totally change dimensions in the coming years. Indeed, immersive cinema was described in 2015 as "the seventh art." This concept has been around for some time, but major success in its use first began in 2015. The goal of immersive cinema is not for the viewer to sit quietly on their sofa watching a movie. It consists of using the movements of the body and the head to explore the surrounding environment. The audience then

3D Printing
ISBN 978-0-323-99463-7
https://doi.org/10.1016/B978-0-323-99463-7.00017-7

becomes the first actor in the film. The wearing of a helmet is not always required to participate in immersive cinema.

The use of large screens, increasingly larger and with a greater circumference, have made it possible to greatly improve the broadcasting and transmitting possibilities of film by causing different results on the retina. The laws of framing have been radically changed, as well as the internal rhythm and montage, which have undergone important transformations, going from an almost square screen to the landscape format, nowadays causing a problem for filmmakers and projectors in adjusting to the various projection formats (Trabucchi & Magistretti, 2020).

The tests, already carried out and used in cinema exhibitions, fairs, or theme parks, combine the new filming and projection systems with the use of spherical screens, Omnimax or similar, totally enveloping the viewer, and also achieving other movement effects through armchairs, sound accompaniments, and other smell and touch effects.

Holography, used in the film industry, provides stunning three-dimensionality in avatars. They are technically very advanced experiments with a lot of aspirations, but still impossible to apply to commercial theaters for the general public. They are still on the stage where the movie started, screened at the fair, like unheard of.

17.3 Why

Stereoscopic procedures, like 3-D, which have not yet achieved a perfect set-up, in the 21st century have experienced a new boom, which the large film companies aiming to compete with other media and refill cinemas with growing audiences (Urban, 2020).

Increasingly, high-quality reproduction systems are available in homes: large screens, digital video reproduction systems, surround sound, and other wonders that enhance the home theater experience. Therefore, it is logical that something different is required to entice viewers to watch in cinemas. There are different systems offering 3-D cinema that have been refined over time, but all are based on the same principle: stereoscopic vision.

The next step seems to be the holographic cinema, still under development, which consists of a screen made up of a multitude of small units whose transparency, brightness, and color can be changed. When the light falls on these elements in a different state, the light reflected on the screen forms the desired image. If the changes are made quickly enough, the sensation of movement is achieved.

Our eyes are like cameras that obtain flat, two-dimensional images. Due to the gap that exists between the eyes, this binocular vision achieves two images that are slightly different, and that difference varies depending on the distance at which the different objects that fall in our field of vision are.

Our brain is in charge of interpreting these flat images in a way that builds the three-dimensionality to which we are accustomed. It is an essential evolutionary adaptation for many animals that, for example, need to move safely by jumping from one branch of a tree to another. Without stereoscopic vision, it is impossible to calculate distances correctly. It is also indispensable for predators, who need to calculate the distance to prey in order to hunt efficiently.

Stereoscopy is a result of the attempt to show a three-dimensional image on a two-dimensional plane, and contrary to what is commonly assumed, stereoscopy preceded photography with stereoscopic drawing. As early as the 15th century, Leonardo da Vinci was already studying this intriguing phenomenon.

For the brain to be able to perceive a third-dimensional image requires data on the distance of the objects; this information is obtained thanks to the fact that we have two eyes, with each perceiving objects from a different angle, resulting in a triangulation from which the brain gets the distance to the object. If we take two images with a slightly different angle (with different parallax) and provide it to each eye, the brain will be able to reconstruct the distance and, therefore, the sensation of depth.

The reconstruction of depth from two images is known as **stereopsis**, and this capacity is not the same in all people. Approximately one in 20 individuals cannot perform it and will not be able to see the 3-D effect. The purpose of the different 3-D or stereoscopic techniques is to provide each eye with a different image.

17.4 When

Photography in three dimensions was born almost with photography itself. In fact, when photography was invented in 1839, enough was known about three-dimensional vision, but it was about 10 years later, around 1850, that the first three-dimensional photographic images were taken and released. A decisive factor in the spread of three-dimensional photography was when Queen Victoria of the United Kingdom, captivated by them, ordered such images to be taken. This marked the beginning of the great popularity of three-dimensional photography in the 19th century, particularly in Europe and the United States.

17.5 Who

The stereoscope is a device that consists of binoculars that allow us to observe reality with a 3-D effect, and that later and together with the discovery of the photographic camera gave rise to the stereoscopic camera (Yekti, 2017).

Its inventor was Charles Wheatstone, an English experimental physicist who was the first to take an image with a 3-D effect, using an apparatus previously invented by him in 1833.

It was used by surveyors to make topographic measurements until satellite procedures appeared, by observing photos taken by airplanes with stereoscopes.

A stereoscopic image is a 3-D image. It is made up of two overlapping images with a minimal difference between them. These two images can be viewed separately with the help of special glasses (3-D glasses). These lenses filter the images by making one eye perceive one image, while the other perceives another image, causing the brain to combine the images, hence giving the sensation of three dimensions.

Stereoscopic vision is based on the real vision of humans, binocular vision (with two eyes) that produces the sensation of a three-dimensional image, as the two images captured by the retinas are processed by the brain at the same time.

Between the 1920 and 1930s, numerous short films were produced in three dimensions, which contributed to research into the techniques, and enhancement of the filming and projection system.

However, the first 3-D talking movie was an Italian film, *Nozze vagabonde* (*Vagabond Weddings*) made in 1936 by Guido Brignone, featuring Leda Gloria, Maurizio D'Ancora, and Ermes Zacconi. The director of photography was Anchise Brizzi.

Pete Smith's *Audioscopiks* (1935), an 8-minute documentary short, was the first film in relief and was nominated for the Oscars.

In 1937, the first color 3-D film, *Zum greifen nah* (1937), was screened, produced in Germany, in which Polaroid filters were used. In 1939, a short film initially in black and white was presented at the New York World Exposition, *In Tune with Tomorrow*, which was so successful that the following year it was remade in color.

17.6 Where

In 1853, Hollman devised a projection in three dimensions on a screen, which was improved on 5 years later in 1858 by Charles D'Almeida, and then by A. Molteni in 1890. From 1891 to 1893, Louis-Arthur Ducos du Hauron experimented with anaglyphs, improving Wheatstone's invention, along with Charles D'Almeida and A. Molteni. Anaglyphs use the projection of two simultaneous images 65 mm apart from each other, with the images being projected by two projectors, one green and the other red, while the viewer wears two-color lenses, also red and green, to finally obtain a single image that is black and white and embossed. These were superseded in 1937—38, by tinted windows experimented on by the German opticians Stterkon and Zeiss, and by the American engineer John Norling.

In 1935, Pierre Couvier and Louis Lumière presented some of their films in relief. Ivanov, in 1938, achieved a relief system without glasses by a modification of the screen.

17.7 How

The large studios, despite the decline of the relief cinema, continued to investigate the method, and after resolving some technical problems they tried to revive the format.

Arch Oboler, who had a restless and creative spirit, created the *Space-Vision 3-D system*, in which the two images that previously had to be reproduced on separate projectors, were superimposed on a single celluloid tape so that a single projector, to which a special lens was inserted, could be used. This procedure was used in the feature film *The Bubble* (1963), about a fantastic invasion of Earth, that was well received by the public but poorly by critics, marking the revival of cinema in relief and the interest of many film studios.

In 1970, the Stereo-Vision brand developed another system in which images were compressed side by side on a single 35 mm film strip, and projected by an anamorphic lens through Polaroid filters. With this system, the danger of desynchronization was eliminated. The format was later repeated in many films and was used to create erotic comedies full of violence and nudity. Stewardess, Al Silliman Jr., starring Christina Hart, Michael Garrett and Paula Erickson, was such a financial success that it worked together in Silliman a few years later reissued in large format in Joe Alves' 1983 Jaws 3-D.

(From: Lechner, C., & Pervaiz, A. (2020). From invention to industry from a social movement perspective: The emergence of the 3D printing industry. Journal of Innovation and Entrepreneurship, *9(1), 1–28.)*

Cinema is constantly reinventing itself to fill theaters, with new technologies proposed and marketed, and often supplemented with special effects. The cinema in three dimensions, which has already emerged at various times in history, is revived with new visual stimuli to compete with the pirating of films and to entice more people go to the cinema. The good news is that new digital techniques are still being researched and applied, and, in good cinematographic hands, great products could be produced. The objective of producers and filmmakers is to offer a spectacular experience that hypnotizes the viewer, and that is a sensation that cannot be replicated on the sofa at home or on the computer screen (Regehly et al., 2020).

The director James Cameron edited in 3-D-IMAX format, in 2003, *Ghosts of the Abyss, Mysteries of the Titanic*, about the legendary sunken ocean liner Titanic and the underwater dives that have revealed its current state on the ocean bed. It was the first feature film to be edited with the "reality camera" system, a very lightweight video camera that could be carried on the shoulder in HDTV format instead of film. *Ghosts of the Abyss* marked the "before and after" in 3-D cinema. Later he also made the documentary *Aliens of the Deep*, inspired by concepts from the field of astrobiology, to improve results, and created new camera models and introduced the possibility of shooting extreme close-ups.

In 2004, *The Polar Express*, an animated film directed by Robert Zemeckis, completely reworked to be enjoyed in three dimensions, is considered the first animation in 3-D-IMAX format feature film. The series was 14 times bigger than the 2D version, rekindling interest in 3D movies.

Interesting films from this stage include *Spy Kids 3-D – Game Over* (2003) by Robert Rodriguez, *Chicken Little* (2005), the first 3-D digital animation film entirely produced by Disney, and *Monster House* (2006), an innovation by Steven Spielberg as a producer in addition to working in collaboration with producers such as Robert Zemeckis, Steve Starkey, and Jack Rapke, the revival of *Nightmare Before Christmas* 3-D (2006), *Superman Returns* (2006), among many others.

However, the real breakthrough in digital projection with the *Chicken Little* (2005), cited above. It is no longer necessary to synchronize two projectors, one for each eye. The first stop-motion animated film planned and shot directly in 3-D was *Coraline,* Henry Selick's *Worlds of Coraline* (2009).

Despite the reconversion of theaters for the new 3-D digital system being very expensive, 3-D cinema opens a new opportunity. Some of the films already shown, such as *Avatar*, have surpassed all box office records. Also, Disney, for example, is preparing new versions of its classics in 3-D.

Significant films are, apart from the aforementioned, include:

Monsters vs. Aliens (2009), by Rob Letterman and Conrad Vernon, from the Dreamworks factory;

Up, by Pete Docter (2009), produced by Disney and Pixar;

Avatar (2009), by James Cameron;

A Christmas Carol (2009), by Robert Zemeckis, produced by Disney;

Alice in Wonderland, (2010), by Tim Burton.

Hubble is a documentary made with an IMAX 3-D camera, which was recorded on board the space shuttle *Atlantis*, featuring spacewalks and the efforts of seven astronauts to repair the Hubble space telescope. With the support of NASA, the production was shot under the orders of Toni Myers, with Leonardo DiCaprio as the narrator.

(From: Mendis, D., Lemley, M., & Rimmer, M. (Eds.). (2019). 3D printing and beyond: Intellectual property and regulation. *Edward Elgar Publishing.)*

17.8 Types of glasses for 3-D

Further details of glasses for 3-D can be found in the article by Khairallah et al. (2020).

17.9 Passive 3-D glasses (anaglyph lenses)

Typical passive 3-D glasses use anaglyphic lenses, which have two different colored lenses (usually red and blue) to filter images from a 3-D movie. The effect was patented by Louis Ducos du Hauron in 1891.

The anaglyph system images are composed of two layers of color that are placed with a slight deviation from each other to create the effect of depth. Looking through anaglyph lenses creates a 3-D image, as each eye sees a color and the brain merges the images into one. When looking without the glasses, a double image in blue and red is seen. The glasses were originally made of cardboard and two-tone cellophane, but the images weren't sharp enough. The best clarity is achieved with molded acrylic lenses, which make up the difference.

References

Khairallah, S. A., Martin, A. A., Lee, J. R., Guss, G., Calta, N. P., Hammons, J. A., Nielsen, M. H., Chaput, K., Schwalbach, E., Shah, M. N., Chapman, M. G., Willey, T. M., Rubenchik, A. M., Anderson, A. T., Wang, Y. M., Matthews, M. J., & King, W. E. (2020). Controlling interdependent meso-nanosecond dynamics and defect generation in metal 3D printing. *Science, 368*(6491), 660–665.

Lechner, C., & Pervaiz, A. (2020). From invention to industry from a social movement perspective: The emergence of the 3D printing industry. *Journal of Innovation and Entrepreneurship, 9*(1), 1–28.

Mendis, D., Lemley, M., & Rimmer, M. (Eds.). (2019). *3D printing and beyond: Intellectual property and regulation*. Edward Elgar Publishing.

Regehly, M., Garmshausen, Y., Reuter, M., König, N. F., Israel, E., Kelly, D. P., … Hecht, S. (2020). Xolography for linear volumetric 3D printing. *Nature, 588*(7839), 620–624.

Trabucchi, D., & Magistretti, S. (2020). The battle of superheroes: The rise of the knowledge platform strategy in the movie industry. *Journal of Knowledge Management, 24*(8).

Urban, P. (September 2020). Graphical 3D printing: Challenges, solutions and applications. In *London imaging meeting* (Vol. 2020, pp. 87–90). Society for Imaging Science and Technology. No. 1.

Yekti, B. (2017). Visual tactility of 3d printing utilization in stop motion animation. *Advanced Science Letters, 23*(1), 126–129.

CHAPTER 18

3-D printing in the tool and die industries

Contents

18.1 How

Mold parts are usually made of tool steel with low workability. The first task of any manufacturing process is to deliver high-quality parts as efficiently as possible. 3-D printing has recently reached an inflection point; while new processes and implementations are shaking up the market, adoption rates are also accelerating. In addition to 3-D printing metal (which in itself is a big step forward), researchers at Oregon State University recently announced a technology that can print expandable electronic devices, foldable displays—even "soft" robots (Berman, 2012).

Many challenges in stamping die production can be directly attributed to the machining process. Through 3-D printing, the use of complex shapes made of hardenable tool steels such as A2 and D2, and working mold parts for cutting, stamping, and bending are a good prospect for metal additive manufacturing. The most common form of large-scale manufacturing of plastic parts is injection molding, which involves injecting liquid plastic into a high-pressure mold.

3D Printing
ISBN 978-0-323-99463-7
https://doi.org/10.1016/B978-0-323-99463-7.00018-9

18.2 Who

Boeing and Rolls-Royce recently invested $37 million in British aerospace company Reaction Engines to develop a 3-D-printed engine. Russia is using additive manufacturing for its next-generation jet engine Aviadvigatel PD-35. Boeing is also working with DARPA and other agencies to develop 3-D printing solutions and standards for aircraft. Automakers including General Motors use 3-D printing for auto parts, and GKN is developing new metal powders for industrial use. At sea and on land, the U.S. Navy plans to have nearly 1000 approved 3-D-printed components for its fleet by the end of 2018. Research and Markets believes that, driven by industrial applications, by 2023, industrial 3-D printing will become a $5.6 billion market' beyond simple prototype manufacturing, such as component manufacturing and machining. Efforts are being made all over the world to develop molds and other industrial devices. There are opportunities for designing disposable and unique components, especially in the consumer market: French appliance manufacturers SEB and Boulanger use this technology to print spare parts on demand. New categories of consumer goods are also possible. Researchers at Carnegie Mellon University used cheap 3-D printers to produce flat objects that fold into a predetermined shape when heated: Will ordinary household items be manufactured and distributed as flat objects in the future (Berman, 2012)?

Die casting is a process very similar to injection molding, except that molten plastic is replaced with molten metal. Liquid metal is injected into the metal mold (in this case the same as the mold) under high pressure, filling the cavity of the mold and hardening to form the metal part. Another aspect to investigate when choosing a 3-D printer is why one should be chosen over the other, especially when there are dozens of different products at different prices. At this point, other people can also come in handy, that is, those who have inspected industrial 3-D printers before, and bought and used them, thus forming an idea of their suitability. Before purchasing a printer, it should be assessed for its suitability, but initially the choices of others can be sought, while choosing the best-selling and most commented on current industrial 3-D printer can be done. It is for this reason that an updated list (monthly) of the best-selling 3-D printers is available on Amazon, which can support the decision-making process.

Without adequate control systems, the food, pharmaceutical, steel, and especially automotive industries cannot function today. Thanks to well-designed and customized tools, this process can be automated; for example, processing errors can be detected and corrected in a timely fashion, thereby preventing defective products that are expensive to replace. Now, with the advanced vision system, everything becomes simpler. Manufacturers that provide industrial automation solutions are increasingly using 3-D printing technology.

18.3 Why

The high labor cost and the concept of Industry 4.0 are closely related to the gradual digitization of the production process, which means that companies are moving away from human control and focusing on full automation and robotics in this field. A vision system that can accurately identify and verify product locations, read barcodes, and detect damaged or incomplete batches on the production line has become a valuable support for this organized production control. Installing CCTV cameras on the test bench is easier and faster than 3-D printing components. Control systems work better than humans in many aspects (but not all): they can resist fatigue, which leads to loss of attention, which is very valuable in this situation. They are ideal for the food or automotive industries, for example, to identify parts for various car models delivered on assembly lines. However, in order to match industrial image-processing systems (sensors and cameras with integrated software) to human abilities and perception capabilities, their suppliers have been working on increasingly advanced methods and improvements. One of the key elements of the tested vision system is the correct installation of CCTV cameras. Their correct positioning is critical to the entire process, and a custom mounting bracket that matches the size of the camera is required (Kang & Ma, 2017).

Before starting to use the 3-DGence printer, we used a universal articulated arm with six degrees of freedom to mount the camera. Such a solution is not only unpleasant, but also takes time. The installation took up to 4 hours, and the operator had to adjust the position of the arm several times. Today, in just a few minutes, we started printing a fully functional and simple 3-D stent. The cameras used in control applications differ in terms of technical parameters, electronics, and most importantly geometric shapes, which is why their mounting brackets must match accurately. The best and cheapest way to customize is 3-D printing technology. The use of 3-D printing technology also significantly reduces the production cost of the parts used in the assembly of a vision system. Components made on site using materials such as ABS are cheaper than components ordered from external suppliers or aluminum (Stansbury & Idacavage, 2016).

(From: Is 3D printed tooling a solution for die casting? (2019). Retrieved February 16, 2021, from https://www.productionmachining.com/articles/is-3-D-printed-tooling-a-solution-for-die-casting.)

18.4 When

Although it has been around since the 1980s, recent advances in technology and materials science, especially metal 3-D printing, have finally shown that it has withdrawn from the niche and consumer markets. Technology is becoming cheaper, faster, and more powerful. Thanks to innovations brought to the market, inks have become more flexible and affordable.

The use of 3-D printing technology has also significantly reduced the production cost of the parts used in the assembly of a vision system. Components made with materials such as ABS on the company site are cheaper than components ordered from external suppliers or aluminum. Companies that are brave enough to invest in advanced 3-D printers will soon notice the real benefits and profits of being able to fully control the geometry of the printed elements. If such a project is outsourced, just because of poor adaptability, you may lose a few days to make the necessary changes, and of course you will have to pay additional start-up costs. Good and precise equipment for industrial image-processing guarantees the highest quality and flexibility in the production process, which is why it is so important for its suppliers to properly match all elements to the customer's individual business and environmental needs. Taking advantage of the opportunities provided by 3-D printing, we can meet these expectations and ensure the visual stability and repeatability of customers in the process of visual quality control. Suppliers can enjoy lower costs, greater flexibility, and a more competitive market (Stansbury & Idacavage, 2016).

18.5 Where

The number of printable materials continues to increase; the most common being thermoplastic polymers, which are used by filament deposition printers, such as ABS, PLA, nylon, polycarbonate, and polystyrene; in photopolymerization technology, acrylic resins and rings are mainly used. Some online printing services also allow you to choose ceramics, precious metals (gold and silver), titanium, brass, and bronze. Professional 3-D printing systems are widely used to create prototypes, so any issues related to styling, adaptability, and function can be aesthetically evaluated and checked. In recent years, we have witnessed the rapid expansion of the use of professional 3-D printers in the production environment for the construction of final parts and the creation of equipment for direct production, thanks to the increased availability of higher-performance additives and materials. In the field of plastic molding, the most widely used 3-D printer is the prototype of the final product, which is used to facilitate the research and construction of metal molds. However, there are opportunities to gain huge advantages in the realization of impressions and injection tools.

18.5.1 Injection molding

Injection molding is a manufacturing process in which molten plastic material is injected into a mold cavity, where it cools and hardens to form the shape of the cavity. It is one of the most popular processes in the world, and is especially suitable for the mass production of high-precision and complex finished parts. By introducing a 3-D printer for a company's production, it is possible to obtain prototypes, molds, and small batches of finished parts made of metal or plastic. Additive manufacturing is part of the third industrial revolution, which began in the 1970s in electronics, information technology, and telecommunications. It is also part of the upcoming digital factory (Industry 4.0) and robotics, nanotechnology, augmented reality, Internet of Things, etc. This is a very important step. Since the advent of additive manufacturing, companies that have used consciously for many years in the most structured and innovative ways are a tool for companies that want to increase productivity, shorten time to market, and save money. This is the case for small batches of plastic products, which can be manufactured with a 3-D printer instead of resorting to injection molding, which requires expensive steel molds for traditional processes. When the number of products is larger, through 3-D manufacturing, a mold can be created in a special polymer with high temperature resistance and then inserted into a traditional molding machine instead of a metal molding machine. For application with industrial 3-D printers, there are many technologies, some are very different, while others differ only in small details.

18.5.2 Direct metal printing (DMP)

DMP was developed by 3-D Systems and allows the use of steel, aluminum, titanium, and cobalt–chromium powders with extremely small particle sizes. Therefore, it is possible to obtain well-defined metal parts suitable for final products and functional prototypes in various fields, from jewelry to industrial-grade additive manufacturing to applications in the dental field. As with the normal sintering of metals used in other processes, the heat that allows powder to solidify layer by layer comes from a laser with a power of 50–500 W. 3-D printers used for DMP production include ProX 350, ProX 500, and ProX 100.

18.5.3 Multijet printing (MJP)

This technology was developed by 3-D Systems, which is the only manufacturer that offers this technology in its Projet series of machines. An inkjet-like nozzle "sprays" layers of wax-like thermoplastic resin layer by layer, which is cured by ultraviolet light. Its main advantage is reflected in the combination of good implementation speed and excellent resolution. There are printing systems suitable for the production of small batches of finished products, prototypes, silicone mold models, or investment casting molds. Some materials can also withstand very high temperatures.

Using 3-D printing in tool and mold manufacturing, there is almost no limit to the production of various shapes. This makes many possibilities, including the production of small batches or customer-specific products and spare parts. Where small batches, or single or special parts are required, 3-D printing shows its advantages in tool and mold manufacturing. Industrial thermoplastics in high-resolution shapes (such as digital ABS) are used to create production parts. Also, conductive plastic is used to make clamping and measuring devices. A soft surface is required to protect sensitive products, etc. (Jahan & El-Mounayri, 2016).

The limitations of traditional methods in tool manufacturing do not apply to 3-D printing in tool and mold manufacturing. In addition, 3-D printing in tool and mold manufacturing significantly improves the cost efficiency and optimizes production processes. An example of using injection mold 3-D printing in tool and mold manufacturing is drawing parts from Ultem1010 3-D printing material. The benefits from 3-D printing include fast and demand-oriented tool production. The manufacture of tools and equipment is one of the most complex areas in the manufacturing process. In traditional methods, this process is time-consuming and expensive, however, through additive manufacturing, different quantities of individual parts or series of products can be manufactured faster and at a lower cost. This means that customer-specific tools generated directly from CAD can be ready in a few days. Costs and risks can be reduced by avoiding costly mistakes by verifying tools, processes, and final products. With 3-D printing in tool and mold making, you can save waste and rework. Since the time required to process and integrate components is minimal, and post-processing is almost completely eliminated, 3-D printing tools can save a great deal of money. Personal adjustments can be made for better ergonomics. Flexibility can be enabled and lightweight and sturdy tools and equipment can be created using 3-D printing. These can be easily adjusted individually for each application and provide better ergonomics and higher user safety. Durable and elastic materials can create a variety of shapes for composite product manufacturing. 3-D-printed cores made of soluble materials also allow the use of composite materials to make hollow and complex parts. After the pad hardens, the core is washed off. There is no longer a need to create each core manually, but instead additional print cores are created from existing CAD files for any number of repetitions. Casting and injection molds shorten the development time and reduce risks at the same time. Also the final production process and materials can be tested to verify molded parts. The injection mold in 3-D printing is used directly in the small batch production of digital ABS. 3-D printing is also very suitable for producing stretch blow molding/injection stretch blow molding (blow molding) molds used to produce hollow plastic bodies. In the casting mode, fast and accurate 3-D casting models are obtained at low cost. Even if the complexity of the model increases, no additional costs will be incurred. The original form of a product can be created directly from the CAD program, the "master mode"—without any intermediate manual steps. The manufacture of tools and components is one of the most complex steps in the manufacturing process. Traditional methods are time-consuming, expensive, and technically demanding. Here, it is usually necessary to weld individual components together, which is not always easy. This is different from additive manufacturing, where, 3-D printing technology is very cost-effective in producing different quantities of individual parts or a series of products at once, but most importantly, it is faster and more flexible.

18.6 Advantages of 3-D printing in tool manufacturing

The advantages of additive manufacturing in tool manufacturing include:

- Improve the quality of manufactured objects;
- Great freedom of design;
- The scrap rate is also significantly reduced;
- The unit cost must be able to withstand international comparisons.

(From: 3-D printing, DMLS for the tooling industry - i3DMFG. (2021). Retrieved February 16, 2021, from https://www.i3dmfg.com/industries/tooling/.)

18.7 What

Shorter production times also ensure that corresponding offers can be sold on the market faster. Experience has shown that the cost of such a system is sometimes very large in its size and quickly amortized. According to experts, good functional integration can be achieved by using 3D printing. At the same time, independent processes reduce labor costs. 3-D printer manufacturers continue to improve their technology. Therefore, it is now possible to integrate cooling channels into tools and systems with the help of the DMLS process. This can improve heat dissipation. It should not be forgotten that additive manufacturing is also very suitable for repairing damaged knife blades. For example, this can be achieved with a 3-D printer using the DMLS process (Lim et al., 2016).

We regularly report on innovations in tool manufacturing in the *3D-limitless* magazine. We mainly focus on large companies dedicated to 3-D printing. However, sometimes we also report on companies that are familiarizing themselves with the technology for the first time or have already done so. Our articles often provide information about institutions that use large-scale 3-D printing systems and are also dedicated to research and development of unique technologies, such as the Fraunhofer Institute. The 3D-limitless newsletter is highly recommended, it covers many areas of 3D printing and additive manufacturing. If one wants to know more about 3D printing, one can find different information directly in the previously published newsletter.

Behind the fascinating term "3-D printing" is a series of additive manufacturing technologies. Most use CAD data equipment to build objects layer by layer, for example, from ceramics, metals, or synthetic resins. The layer construction method has two basic advantages—first is the principle of free complexity. Although the complex shapes used cause engineers a headache because their structure must be designed to be suitable for production, the complexity of the layering process is not important. Second, economies of scale are a thing of the past. In traditional production logic, the cost of a single project was high. The product will only provide financial rewards after thousands were sold. 3-D printing is now rewriting these rules.

References

Berman, B. (2012). 3-D printing: The new industrial revolution. *Business Horizons, 55*(2), 155–162.

3D printing, DMLS for the tooling industry. (2021). i3DMFG. Retrieved 16 February 2021, from https://www.i3dmfg.com/industries/tooling/.

Jahan, S. A., & El-Mounayri, H. (2016). Optimal conformal cooling channels in 3D printed dies for plastic injection molding. *Procedia Manufacturing, 5*, 888–900.

Kang, J. W., & Ma, Q. X. (2017). The role and impact of 3D printing technologies in casting. *China Foundry, 14*(3), 157–168.

Lim, C. W. J., Le, K. Q., Lu, Q., & Wong, C. H. (2016). An overview of 3-D printing in manufacturing, aerospace, and automotive industries. *IEEE Potentials, 35*(4), 18–22.

Is 3D printed tooling a solution for die casting?.(2019). Retrieved 16 February 2021, from https://www.productionmachining.com/articles/is-3d-printed-tooling-a-solution-for-die-casting.

Stansbury, J. W., & Idacavage, M. J. (2016). 3D printing with polymers: Challenges among expanding options and opportunities. *Dental Materials, 32*(1), 54–64.

CHAPTER 19

3-D printing in the toy industry

Contents

19.1 What

Since entering the new millennium, the concept of "3-D" has firmly entered our daily lives. First, it was associated with film photography or animation. But now, almost everyone has heard of the novelty as 3-D printing. What is it, and what new opportunities in creativity, science, technology, and daily life does 3-D printing technology in the toy industry bring us? This is what we discuss in the following chapter. But first, we start with a bit of history. Although there has only been a lot of discussion about 3-D printing in recent years, in fact, this technology has been around for a long time. In 1984, Charles Hull developed 3-D printing technology to use digital data to replicate objects. Two years later, he named and patented the stereolithography technology. At the same time, 3D Systems developed and created the first industrial 3-D printer. Subsequently, the baton was taken over by 3-D Systems, which in 1988 developed the SLA-250 printer model for 3-D printing at home. In the same year, Scott Grump invented the fused deposition model. After several years of relative calm, in 1991, Helisys developed and began the sale of technology for the production of multi-layer objects. A year later, in 1992, DTM introduced the first selective laser welding system. Then, in 1993, Solidscape was established and began mass production of inkjet printers, capable of producing small parts with ideal surfaces at a relatively low cost. At the same time, the University of

3D Printing
ISBN 978-0-323-99463-7
https://doi.org/10.1016/B978-0-323-99463-7.00019-0

Massachusetts applied for a patent for a 3-D printing technology similar to the traditional 2-D printer inkjet technology. However, perhaps, the peak of the development and popularization of 3-D printing is still to come in the 21st century. In 2005, the first 3-D printer capable of color printing appeared. This was the idea of Z Corp., and was called the Spectrum Z510. Two years later, the first printer capable of copying 50% of its own components appeared. At present, the application scope and application fields of 3-D printing are constantly expanding. It has been proven that these technologies are applicable to everything from blood vessels to coral reefs and furniture. However, we discuss the application areas of these technologies later.

The toy world is no exception. Mid-market companies prefer to use 3D printing. The more democratized the technology, the more it can find new applications. This includes the production of finished 3-D toys and other items that children can use. For example, the industrial giant Mattel recently proposed a 3-D printer called "ThingMaker" for printing polyamide objects. Parts are printed very quickly (averaging half an hour), and the tablet application allows predefined models to be downloaded. Children can now learn about additive manufacturing while making toys.

Other companies are using additive manufacturing to provide another form of entertainment for young children. For example, the Makielab company, produces dolls that are slightly different from the usual dolls: these small dolls represent people with disabilities. Many of us feel nostalgic for our favorite toys from our childhood, because toys have a special place in each child's memory, after all, they have an impact on the consciousness of us at our most susceptible age. Like children today, toys are developing rapidly. To ensure continued demand for toys, they must be:

- Unique;
- Fashionable;
- Cost-effective;
- Fun and safe.

19.2 When

Until recently, traditional design and prototyping methods were complex, expensive, and time-consuming. With the advent of 3-D printers, the development process has been simplified and greatly speeded up. Although different materials can be used to make models, today, plastic-based polymers are most commonly used, which presents a huge opportunity.

Compared with other technologies previously used, 3-D printing is a fairly versatile technology. It allows engineers to use their creativity and versatility on a large scale. This is what toy manufacturers have been waiting for. The real possibility of 3-D printing began to emerge in the era of personal computers, when projects began to be executed in digital form, which made the process of creating models infinitely diverse. It provides an excellent opportunity to produce multiple versions of samples of the same design but with different ergonomics and design elements. The convenience of this technology also provides manufacturers with the opportunity to quickly test copies and show them to experts, potential buyers, or distributors.

Over the past few years, new changes have taken place in the toy market: it is more ethical and ecological. In 2009, when the French Association found through research that a large number of toys contained toxic substances such as phthalates (endocrine disruptors) or formaldehyde, the European Commission formulated a new directive to improve the safety of toys and comply with the threshold of CMR (carcinogenic, mutagenic, or toxic to reproduction) substances and heavy metals such as lead or mercury. Then, a small number of consumers turned to safer, and more ethical or ecological toys. A few years later, the lifespan of toys, like the life cycle of many other consumer products, became a point of contention because they were often too short and become unpopular. The toy industry is seriously diluted and in urgent need of modernization. Some companies subsequently decided to design toys that could be repaired or even recycled. In this area, 3-D printing provides a solution that could change the industry, modernize it, and even enable it to "go green" (León-Cabezas et al., 2017).

19.3 Where

Like many other parts, tools, and components, toys are printable. There are many 3-D files for toys, board games, and even figurines on online platforms such as Cults, Thingiverse, or MyMinifactory. Fans can download a large number of 3-D files for printing. Technology also provides them with the opportunity to design models. This can be a source of enjoyment for game lovers. Nevertheless, the personal consumer goods 3-D printing market is still struggling to find its audience in France. However, another purpose is under development, namely repair (Fig. 19.1).

Figure 19.1 McDonald's kid toys. *(From: Matter of Form: Experience Design for Timeless Brands, 2021 (online). Available at https://www.matterofform.com (Accessed 24 March 2021).)*

19.4 Who

In 2019, 3-D printer manufacturer Dagoma launched an operation called "Toy Rescue." This followed a simple observation: "More than 40 million toys are thrown away by Americans each year. In most cases, they cannot be repaired, one of the reasons is: spare parts do not exist." The purpose of the operation was to replace damaged parts: a new leg, a new wheel, or anything that may be missing when repairing toys or completing a board game. "This project first identified which parts of popular toys have been lost or damaged the most over the past 40 years. A team of 3-D designers used a 3-D scanner to model and recreate many parts. These spare parts documents have been posted on the operation's website: Toy-Rescue.com, and are available to everyone (León-Cabezas et al., 2017).

3-D printing technology is a combination of digital technology, laser technology, and new material technology. Adults, including kindergarten teachers, design and manufacture them safely and conveniently according to the individual needs of children. The development of children and the need for curriculum teaching also enable children and other adults, especially parents, to participate in and make their own toys. Colored toys can also allow the toy industry to produce toys that are more in line with children's individual needs while taking into account profits. Starting from the analysis of the problems existing in children's toys, based on the analysis of the root causes of the problems, innovative suggestions for using 3-D printing technology to make children's toys have been put forward (Michalitsi-Psarrou et al., 2019).

After the emergence of 3-D printing technology, it was quickly applied in the field of education. For example, Stratasys, a well-known 3-D printer manufacturer in the United States, announced that 40% of its products are sold to educational institutions. The Ministry of Education initiated a project to promote 3-D printing technology in STEM (science, technology, engineering, and mathematics) teaching and product design. More and more educators are recognizing the great value of 3-D printing technology in the field of education, while increasing numbers of 3-D printer manufacturing companies have designed printers and software specifically for educators and students. Similarly, 3-D printing technology has gradually impacted the development and manufacture of children's toys. Some developed countries have begun to put forward specific requirements for kindergarten teachers to master and use 3-D printing technology, and have also conducted special training for kindergarten teachers (Fig. 19.2).

19.5 How

Many domestic toy manufacturers still use traditional production methods for production. The process is roughly as follows: conception → manual drawing of floor plan → computer software drawing of three-dimensional drawing → prototype toy parts → assembly verification → rework → re-verification. Practice has proved that the design process can cause great

Figure 19.2 3D Models for Professionals: TurboSquid. *(Drink, F., Design, I., Models, F., Models, 3., Models, M., Models, C., Models, B., Models, L., Models, A., Models, R., Models, O. and Models, F. (2021).* 3D models for professionals. *TurboSquid. (online). Available at: https://www.turbosquid.com. (Accessed 24 March 2021).)*

wastage of manpower and material resources. However, the application of 3-D printing technology in the toy industry is mainly manifested in the following aspects (Chen, 2017). (1) It shortens the product development cycle: without machining or any molds, parts of any shape can be generated directly from computer graphics data, which greatly shortens the product development cycle, improves production efficiency, and reduces production costs, which is extremely beneficial to the improvement of company competitiveness. (2) Toy customization is easier: Toy customization or a high degree of personalization is already easy to achieve. (3) New toy product development: It can realize some very complicated structures and machinery, develop toy forms that cannot be completed by traditional manufacturing methods, and bring new vitality and profit growth points to the toy industry. (4) A new toy sales model is possible: With 3-D printing technology, toy manufacturers can provide 3-D drawings instead of selling physical objects to customers, so that customers can print out the toys they are interested in at home. Experience can also lower the purchase price; this toy sales model is also more environmentally friendly because it reduces logistics and warehousing, and reduces carbon emissions, which is the future trend. The emergence of 3-D printing based on mainstream technology in the toy industry helps companies quickly develop iterative products; realize complex prototype designs that cannot be completed by traditional manufacturing models; new product designs do not need to be outsourced; and there is no need to worry about information leakage. So what is the current application situation of 3-D printing in the toy industry? In this regard, the representative of Stratasys believes (Nur-A-Tomal et al., 2020) that toy companies use different materials or different models according to the different functions of the products they want to test. For example, FDM and ABS or PC materials are used for locking and assembly testing. In addition, they will use the latest full-color equipment to make color multimaterial prototypes to test the proportions and coordination of colors and the softness of components. Some emerging game companies also use lighters or desktop 3-D printers to customize toys, such as decorations with names or special symbols. This type of company may not be as large as a traditional game company, but it promotes another emerging trend and business. In order to promote the application of 3-D printing in the toy industry, Stratasys held meetings with different toy manufacturing associations, held cross-industry application seminars and speeches, and participated in toy fairs with partners (for example, it participated in the 29th Guangzhou Toy Fair in early April 2018) to introduce the current applications and trends of 3-D printing in the industry to different companies. A representative of Stratasys confirmed that their current gaming partners will integrate a variety of 3-D printing technologies and conduct comprehensive tests on new products under development. "Take Hong Kong toy manufacturer Toy State International Ltd. as an example. They use plastic to make early design models and functional prototypes for performance testing. For surface details that require complex conceptual models and small tools, the company's design team will use PolyJet printer prototypes (PolyJet 3-D printing technology is similar to inkjet printing, but instead of ejecting ink droplets onto paper, it ejects liquid photosensitive resin material onto the build tray and cures it with ultraviolet light.) But for certain thin-walled and thin-walled toys, repeated performance testing requires harder materials (such as drone frames). The design team will use FDM equipment and PC-ABS (engineering plastics) to print out models with the same hardness and toughness (Quan et al., 2020)." "Also, some toy companies use full-color 3-D printers to make colorful toys, test the color contrast of baby toys, and mix materials to compare the Shore hardness of different parts of 3-D printed simulation toys."

19.6 3-D printing still faces many technical challenges

Many domestic toy manufacturers still use traditional production methods.

However, the application of 3-D printing technology in the toy industry is mainly manifested in the following areas (Chen, 2017):

(1) Demand product development cycle: Without mechanical processing or any molds, any direct-shaped parts can be generated from computer graphics, thereby greatly improving the product development cycle, improving data production efficiency, and reducing production costs.

(2) Toy customization is easier: Toy customization or a high degree of personalization is already easy to achieve.

(3) Development of new toy products: It is possible to realize some very complex structures and machinery, develop toy forms that cannot be completed by traditional manufacturing methods, and bring new vitality and growth points to the toy industry.

(4) A new toy sales model becomes possible: With 3D printing technology, a new model of toy sales is possible, and toy manufacturers can even offer 3D printing instead of showing toy manufacturing to customers. Customers can print out these items at home and experience it for themselves. It can reduce the purchase price of this toy sales models, because it reduces logistics and transportation, and is also more environmentally friendly and reduces carbon emissions, which is the future development trend.

Based on the emergence of mainstream technology in the toy industry, it is helpful for companies to print, test, and develop products; to achieve prototype designs that cannot be completed by traditional manufacturing models; with new product designs do not need to be so without worrying about information. How's it going? Here, representatives of Stratasys believe (Nur-A-Tomal et al., 2020).

19.6.1 Multiple testing techniques

A representative of Stratasys commented that their current game partners will comprehensively test new products they are developing with a variety of 3-D printing technologies: "Take Hong Kong toy manufacturer Toy State International Ltd. as an example. They used plastic to make early design models and functional prototypes for performance testing. They sprayed the reusable wall resin material onto the material and cured it with light. Equipment - ABS (plastic) prints models with good characteristics and photo-curing 3D printing technology (Quan et al., 2020).

References

Chen, Z. (2017). The service-oriented manufacturing mode based on 3D printing: A case of personalized toy. *Procedia Engineering, 174*, 1315–1322.

Drink, F., Design, I., Models, F., Models, 3., Models, M., Models, C., Models, B., Models, L., Models, A., Models, R., Models, O., & Models, F. (2021). *3D models for professionals*. TurboSquid. (online) . Available at https://www.turbosquid.com/ (Accessed 24 March 2021).

León-Cabezas, M. A., Martínez-García, A., & Varela-Gandía, F. J. (2017). Innovative functionalized monofilaments for 3D printing using fused deposition modeling for the toy industry. *Procedia Manufacturing, 13*, 738–745.

Matterofform.com. (2021). *Matter of form: Experience design for timeless Brands*. (online) Available at https://www.matterofform.com (Accessed 24 March 2021).

Michalitsi-Psarrou, A., Koussouris, S., Kontzinos, C., Markaki, O., Ntanos, C., Panopoulos, D., & &Psarras, J. (2019). Empowering product Co-creation approaches through business interoperability concepts: The toy industry case. In *Enterprise interoperability VIII* (pp. 397–408). Cham: Springer.

Nur-A-Tomal, M. S., Pahlevani, F., & Sahajwalla, V. (2020). Direct transformation of waste children's toys to high quality products using 3D printing: A waste-to-wealth and sustainable approach. *Journal of Cleaner Production, 267*, 122188.

Quan, H., Zhang, T., Xu, H., Luo, S., Nie, J., & Zhu, X. (2020). Photo-curing 3D printing technique and its challenges. *Bioactive Materials, 5*(1), 110–115.

CHAPTER 20

Summary and conclusion

Contents

20.1 Summary

3-D printing technology has gradually begun to occupy niches in many industries, although its process is complicated and it is affected by many factors. The most important of these factors is that the technology cannot currently work in harsh environments. Today, 3-D printers cannot operate in high or low temperatures, high humidity, or strong winds. In addition, an important problem is the limitations of the size of the printer, and the limitations of the building and structure. Many companies around the world are working hard to solve these problems. Perhaps, over time, there will be some significant changes, and 3-D printing technology will become a part of our daily lives.

The use of 3-D printing technology has also greatly reduced the cost of developing new models. For the emerging Asian market, 3-D printing technology also reduces the difficulty in producing newer and more complex products. In the future, the success or failure of many manufacturing plants will depend more on 3-D product design and electronic drawing technology. These are the prerequisites for making better products based on the 3-D printing technology platform.

20.2 Conclusion

Many designers are very interested in the phenomenon of 3-D printing technology, which is not surprising when the infinite creation possibilities it offers are realized. From the most basic forms to the most outlandish, almost anything is possible, thanks to 3-D printing/additive manufacturing. This is

3D Printing
ISBN 978-0-323-99463-7
https://doi.org/10.1016/B978-0-323-99463-7.00020-7

why increasing numbers of designers use 3-D printing in their work. Therefore, the observations are clear: additive manufacturing will revolutionize many industries, especially those mentioned in this book.

3-D printing has both advantages and disadvantages. However, as has been seen, the continuous development and improvement of 3-D printing tools in many areas has been incredible, with 3-D printing technology having a bright future in the hands of designers and enthusiasts who rely on 3-D equipment to develop the most innovative of ideas and projects.

3D printing/additive manufacturing technology undoubtedly has great potential in many areas: body parts or small complex parts, and parts with internal cavities. However, its widespread use of chemical industry products (despite the assurances of Elon Musk, IoT and hip economics defenders) is currently limited by objective flaws in the technology. So far, 3D printing/additive manufacturing has some clear advantages over traditional production techniques.

Ensuring the inertness of the reagents is important when considering the possibility of releasing unwanted impurities into the reaction medium. For certain types of 3-D printing (such as powder sintering), porosity and air permeability need to be considered. The surface quality of key products is an important criterion, which affects the flow of matter in many ways, for example, generating larger turbulent vortices in the boundary layer than machined or cast parts.

There is no doubt that 3-D printing technology will continue to contribute to our lives in the future. The use of new 3-D printing technology largely depends on the operating conditions of the design components and the needs of designers. The performance of 3-D printers has improved continually. In the near future, many of the products we use will be manufactured by 3-D printers and 3-D printing/additive manufacturing will become increasingly popular, with its demand driven by cost reductions and quality improvements.

Index

Note: 'Page numbers followed by "*f*" indicate figures.'

E

F

G

H

Kolibri 360 GmbH
Ferdinand-Jühlke-Straße 7
99095 Erfurt, Deutschland
produktsicherheit@kolibri360.de